Clemens Glombitza

New Zealand coals - A potential feedstock for deep microbial life

Clemens Glombitza

New Zealand coals - A potential feedstock for deep microbial life

Release of potential substrates for the Deep Biosphere during geological maturation and structural evolution of coals

Südwestdeutscher Verlag für Hochschulschriften

Impressum/Imprint (nur für Deutschland/only for Germany)
Bibliografische Information der Deutschen Nationalbibliothek: Die Deutsche Nationalbibliothek verzeichnet diese Publikation in der Deutschen Nationalbibliografie; detaillierte bibliografische Daten sind im Internet über http://dnb.d-nb.de abrufbar.
Alle in diesem Buch genannten Marken und Produktnamen unterliegen warenzeichen-, marken- oder patentrechtlichem Schutz bzw. sind Warenzeichen oder eingetragene Warenzeichen der jeweiligen Inhaber. Die Wiedergabe von Marken, Produktnamen, Gebrauchsnamen, Handelsnamen, Warenbezeichnungen u.s.w. in diesem Werk berechtigt auch ohne besondere Kennzeichnung nicht zu der Annahme, dass solche Namen im Sinne der Warenzeichen- und Markenschutzgesetzgebung als frei zu betrachten wären und daher von jedermann benutzt werden dürften.

Verlag: Südwestdeutscher Verlag für Hochschulschriften GmbH & Co. KG
Dudweiler Landstr. 99, 66123 Saarbrücken, Deutschland
Telefon +49 681 37 20 271-1, Telefax +49 681 37 20 271-0
Email: info@svh-verlag.de

Approved by: Berlin, TU, Diss., 2010

Herstellung in Deutschland:
Schaltungsdienst Lange o.H.G., Berlin
Books on Demand GmbH, Norderstedt
Reha GmbH, Saarbrücken
Amazon Distribution GmbH, Leipzig
ISBN: 978-3-8381-2931-0

Imprint (only for USA, GB)
Bibliographic information published by the Deutsche Nationalbibliothek: The Deutsche Nationalbibliothek lists this publication in the Deutsche Nationalbibliografie; detailed bibliographic data are available in the Internet at http://dnb.d-nb.de.
Any brand names and product names mentioned in this book are subject to trademark, brand or patent protection and are trademarks or registered trademarks of their respective holders. The use of brand names, product names, common names, trade names, product descriptions etc. even without a particular marking in this works is in no way to be construed to mean that such names may be regarded as unrestricted in respect of trademark and brand protection legislation and could thus be used by anyone.

Publisher: Südwestdeutscher Verlag für Hochschulschriften GmbH & Co. KG
Dudweiler Landstr. 99, 66123 Saarbrücken, Germany
Phone +49 681 37 20 271-1, Fax +49 681 37 20 271-0
Email: info@svh-verlag.de

Printed in the U.S.A.
Printed in the U.K. by (see last page)
ISBN: 978-3-8381-2931-0

Copyright © 2011 by the author and Südwestdeutscher Verlag für Hochschulschriften GmbH & Co. KG and licensors
All rights reserved. Saarbrücken 2011

Abstract

During the last decades of biogeochemical and microbiological research, the widespread occurrence of microorganisms was demonstrated in deep marine and terrestrial sediments. With this discovery inevitably, the question of potential carbon and energy sources for this deep subsurface microbial life arises. In sedimentary systems, such a source is provided by the buried organic matter deposited over geological times. During geochemical and geothermal maturation, these organic material undergoes biotic and abiotic alteration processes and is suggested, to release potential substrates into the surrounding. Previous studies showed, that especially oxygen containing compounds are lost from the macromolecular matrix during diagenesis and early catagenesis (Putschew et al., 1998; Lis et al., 2005; Petersen et al., 2008). Oxygen containing low molecular weight organic acids (LMWOAs) such as formate, acetate and oxalate represent important substrates for microbial metabolism. Thus, lithologies containing accumulated sedimentary organic matter (e.g. lignites and coals) may provide a large feedstock for deep microbial life releasing LMWOAs into the pore water during maturation.

In this thesis, lignite and coal samples from sedimentary basins of New Zealand, covering a broad and almost continuous maturity range from Cretaceous to Pleistocene age and representing diagenetic to catagenetic coalification levels (vitrinite reflection: 0.27% to 0.80%), were investigated to estimate their feedstock potential for deep microbial life using a novel developed analytical procedure to analyse kerogen-bound LMWOAs liberated by selective chemical degradation reactions. Formate, acetate and oxalate were found to decrease continuously from early diagenesis to early catagenesis.

This suggests a constant release of these compounds during this maturation interval providing a suitable feedstock for microbial ecosystems in geological time spans. Investigaton of a transect from organic carbon rich (lignite) into poor lithologies (silt and sandstones) from the DEBITS-1 well, drilled in the Waikato coal area (North Island of New Zealand) suggested that the lignite layers sustain microbial communities inhabiting the adjacent more permeable layers (sandstone) by releasing substrates into the surrounding lithologies (*feeder-carrier hypothesis*). Investigations of kerogen-bound high molecular weight fatty acids show for the long chain fatty acids (C_{20}-C_{30}) representing a terrestrial plant material signal a constant decrease during diagenesis and early catagenesis. In contrast, the short chain fatty acids (mainly C_{16} and C_{18}) show an increase again during early catagenesis, an interval where the release of hydrocarbons slowly starts but temperatures are still compatible with microbial life. These increasing abundance of C_{16} and C_{18} fatty acids (being also main constituents in cell membrane phospholipids of bacteria) might suggest an increased microbial community stimulated by the geothermal release of hydrocarbons (*bio-geo coupling hypothesis*).

Hydrolysis of kerogen-bound LMWOAs is suggested to be the most favorable abiotic process releasing potential substrates into the pore water. Kinetic investigations suggested that the hydrolysis appears to be a relatively fast process and, thus, the observed substrate release from lignites and coals over geological times must be influenced and slowed down by further processes such as e.g. pore space, permeability, pore water flow and diffusion. The calculated kinetic parameters (reaction rate constant k', activation energy E_A and frequency factor *A*) point to structural alteration within the macromolecular network during maturation leading to more sterically protected kerogen-bound LMWOAs and, therefore, to a slower substrate release with ongoing maturation.

Additional information about the structure of the macromolecular network were obtained by selective ether cleavage procedure revealing that aliphatic alcohols with more than one hydroxy groups represent important cross-linkage structures. In contrast to the terminal ether-bound mono-alcohols which show a rapid decrease during diagenetic alteration, these compounds show relatively high concentrations even in the more mature coals suggesting that these cross-link bridges are sterically protected within the network structure.

Kurzfassung

Im Verlauf der letzten Jahrzehnte wurde in der biogeochemischen Forschung die Existenz und weite Verbreitung von Mikroorgansimen in terrestrischen und marinen Sedimenten nachgewiesen. Mit dieser Enteckung entstand zugleich die Frage nach möglichen Nahrungs- und Energiequellen für dieses, tief unter der Erdoberfläche existierende, mikrobielle Leben. In sedimentären Becken wird organisches Material über geologische Zeiträume abgelagert. Während der geochemischen und geothermischen Reifung durchläuft das organische Material diverse biotische und abiotische Veränderungen, durch die potentielle Nährstoffe an die nähere Umgebung abgegeben werden können. In vorangegangenen Studien konnte bereits gezeigt werden, dass das makromolekulare organische Material während den Phasen der Diagenese und frühen Katagenese vor allem sauerstoffhaltige Komponenten verliert (Putschew et al., 1998; Lis et al., 2005; Petersen et al., 2008). Solche Komponenten können u.a. niedermolekulare organische Säuren, wie Format, Acetat oder Oxalat sein, die wichtige Nährstoffe für den mikrobiellen Stoffwechsel darstellen. Desshalb können Lithologien die akkumuliertes organisches Material beinhalten (z.B. Lignite und Kohlen) große Vorratsspeicher an Nährstoffen für das tiefe mikrobielle Leben bilden, die niedermolekulare organische Säuren während ihres geologischen Reifungsprozesses an die nähere Umgebung abgeben können.

In dieser Arbeit wurden Lignite und Kohlen aus verschiedenen sedimentären Becken Neuseelands im Hinblick auf ihr Nährstoffpotential für die tiefe Biosphäre untersucht, die eine nahezu kontinuierliche Alterssequenz zwischen Kreide und Pleistozän darstellen und sich im Reifestadium zwischen Diagenese und Katagenese befinden (Vitrinit Reflektion: 0.27% bis 0.80%). Dazu wurde eine neue analytische Methode entwickelt, die es ermöglicht niedermolekulare Säuren aus den Produktgemischen nach alkalischer Hydrolyse (Verseifung) von Lignit- und Kohlenproben zu quantifizieren. Es konnte gezeigt werden, dass diese Lignite und Kohlen im Verlaufe ihrer Reifung kontinuierlich Format, Acetat und Oxalat in Größenordnungen abgeben, die geeignet sind mikrobielle Ökosysteme über geologische Zeiträume mit Nährstoffen zu versorgen. Untersuchungen eines Überganges von organisch reichen Schichten (Lignit) zu organisch armen Schichten (Sandstein) aus dem DEBITS-1 Bohrkern (erbohrt im Waikato Kohlegebiet auf der Nordinsel Neuseelands) legen den Schluss nahe, dass Mikroben in den gröberen, durchlässigeren Sandsteinen beheimatet sind, die in unmittelbarer Nähe der nährstoffreichen Lignit- und Kohleschichten liegen und möglicherweise von diesen mit Substraten versorgt werden (*Feeder-Carrier Hypothese*).

Die Konzentrationen von kerogengebundenen höhermolekularen Fettsäuren, deren Ursprung terrestrisches Pflanzenmaterial ist (n-C_{20}-n-C_{30}), nehmen ebenfalls im Verlaufe der Reifung des organischen Materials ab. Im Gegensatz dazu zeigen die kurzkettigen Fettsäuren (hauptsächlich n-C_{16} und n-C_{18}) einen Konzentrationsanstieg während der beginnenden Katagenese. In diesem Reifestadium beginnt das organische Material, bei Temperaturen die für mikrobielles Leben noch verträglich sind, langsam freie Kolenwasserstoffe zu generieren. Der Anstieg in den Konzentrationen der n-C_{16} und n-C_{18} Fettsäuren (die u.a. wichtige Bestandteile von mikrobiellen Zellmembranen darstellen) kann ein Hinweis sein, dass mikrobielle Gemeinschaften auch innerhalb der nährstoffreichen Lignit- und Kohlelagen existieren, die durch die geothermische Generierung von Kohlenwasserstoffen stimuliert werden (*Bio-Geo Kopplung*).

Hydrolyse ist ein Prozess, der die kerogengebundenen organischen Säuren aus der organischen Matrix freisetzten kann. Untersuchungen der Kinetik dieses Prozesses an Ligniten und Kohlen zeigten, dass der Hydrolyseprozess ein sehr schneller Prozess ist, wesshalb die beobachteten Freisetzungsraten von Substraten in den Sedimenten durch zusätzliche Faktoren, wie z.B. der Porenraum, die Permeabilität des Sedimentes, der Porenwasserfluss oder etwa Diffusion, beeinflusst und verlangsamt sein müssen. Die experiementell bestimmten kinetischen Parameter (Geschwindigkeitskonstante k, Aktivierungsenergie E_A, Arrheniusfaktor A) deuten auf strukturelle Veränderungen im markomolekularen Netzwerk der Lignite und Kohlen hin, die mit zunehmender Reife des Materials zu stärker sterisch abgeschirmten, kerogengebundenen Säuren führen und damit die Freisetzung der Substrate verlangsamen.

Zusätzliche Informationen über die innere Struktur der Matrix konnten durch selektive Etherspaltungsreaktionen erhalten werden. Sie zeigen, das aliphatische Alkohole mit mehreren Hydroxylgruppen eine wichtige Vernetzungsfunktion bilden, die im Gegensatz zu den terminal gebundenen Monoalkoholen weniger stark durch Reifungsprozesse beeinflusst sind und auch in reiferen Kohlen noch in hohen Konzentrationen gefunden werden können. Dies weisst darauf hin, dass solche Netzwerkbildner innerhalb der markomolekularen Struktur der Kohlen sterisch gut abgeschirmt sind.

Contents

Abstract   1

Kurzfassung   5

Abbreviations   29

1. Introduction   33

2. Background   39
    2.1. Microorganisms and the Deep Biosphere 39
       2.1.1. The world of microbes 39
       2.1.2. Discovering the Deep Biosphere 41
       2.1.3. Everything is *not* everywhere 43
       2.1.4. Feeding of the Deep Biosphere - driving Earth's biogeochemical cycles 45
    2.2. Sedimentary macromolecular organic matter - composition, classification and maturation 49
       2.2.1. Kerogen - dispersed macromolecular organic matter . . . 50
       2.2.2. Coal - accumulated terrestrial macromolecular organic matter 65
    2.3. Methods to elucidate the structural information of macromolecular organic matter 76
       2.3.1. The structural composition of macromolecular organic matter as a source of information for organic geochemists   76
       2.3.2. Spectroscopic methods 77

    2.3.3. Pyrolysis 85

    2.3.4. Chemical degradation 88

3. Sample Material And Study Area   **95**

  3.1. Sample material 95

    3.1.1. The coal mine sample set 95

    3.1.2. The DEBITS-1 well 99

  3.2. Geology of the sedimentary basins of New Zealand 103

    3.2.1. Northland Basin 104

    3.2.2. Taranaki Basin 105

    3.2.3. Waikato Basin 106

    3.2.4. West Coast Basin 107

    3.2.5. Eastern Southland Basin 108

4. Methods   **111**

  4.1. Sample preparation 112

    4.1.1. Synthesis of fatty acid ethyl esters (FAEE) 112

    4.1.2. Core subsampling, freeze drying and sample grinding . . 113

    4.1.3. Solvent extraction 113

    4.1.4. Ester cleavage procedures 114

    4.1.5. Ether cleavage procedure 117

  4.2. Instrumental analysis 117

    4.2.1. Open system pyrolysis (pyrolysis-GC-FID) 117

    4.2.2. Gas chromatography-mass spectrometry (GC-MS) 118

    4.2.3. Ion chromatography (IC) 118

    4.2.4. Laboratory equipment 119

  4.3. Software 119

5. Ester Bound Low Molecular Weight Organic Acids Linked To The Kerogen   **121**

  5.1. Introduction 121

  5.2. Sample material 124

5.3. Experimental approach 126
    5.3.1. Sample preparation 126
    5.3.2. Ion chromatography 128
    5.3.3. Method evaluation 129
5.4. Results and discussion 129
    5.4.1. Evaluation of the analytical method for the recovery of low molecular weight organic acids 129
    5.4.2. LMW organic acids linked to NZ coals of different maturity 133
    5.4.3. Feedstock potential of the NZ coals of different maturity 136
    5.4.4. Mechanisms for the release of LMW organic acids from the coal matrix 140
5.5. Conclusion 142
5.6. Aknowledgements 142

6. Ester-Bound Fatty Acids And Alcohols Linked To The Kerogen 145
    6.1. Introduction 145
    6.2. Sample material 147
    6.3. Methods 149
        6.3.1. Sample preparation / saponification 149
        6.3.2. Gas chromatography-mass spectrometry (GC-MS) 152
        6.3.3. Open system pyrolysis GC 153
    6.4. Results and discussion 153
        6.4.1. Alcohols and fatty acids cleaved from NZ coal samples . . 153
        6.4.2. Carbon preference index of short and long chain fatty acids 160
        6.4.3. Potential sources of short chain fatty acids during early catagenesis 163
        6.4.4. Open system pyrolysis of the NZ coal samples 166
    6.5. Conclusion 168
    6.6. Acknowledgements 170

6.7. Additional section: Additional compounds identified in the neutral polar fraction 170

7. Free Fatty Acids, Alcohols And Esters From The Bitumen Fraction   175
7.1. Introduction 175
7.2. Materials and methods 178
     7.2.1. Sample material 178
     7.2.2. Sample preparation 178
7.3. Results and discussion 180
     7.3.1. Fatty acid fraction 180
     7.3.2. Neutral fraction (alcohols, esters and functionalised pentacyclic triterpenoids) 185
     7.3.3. Comparison of free and kerogen bound n-fatty acids, alcohols and functionalised terpenoids 192
7.4. Conclusion 200

8. Organic Acids From Organic Matter Rich Layers From The DEBITS-1 Well   203
8.1. Introduction 203
8.2. Samples and methods 205
     8.2.1. Sample material 205
     8.2.2. Sample preparation 206
8.3. Results and discussion 206
     8.3.1. Low molecular weight organic acids from lignite and coal layers of the DEBITS-1 well 207
     8.3.2. Fatty acids and alcohols from lignite and coal layers of the DEBITS-1 well 210
     8.3.3. Feeding potential of organic carbon rich lithologies for microbial life in adjacent organic carbon poor layers from the DEBITS-1 well, subcore 9.2 (25.93-26.15 m) 214
8.4. Conclusion 220

9. **Kinetics Of The Hydrolysis Of Low Molecular Weight Organic Acids**   **223**
   9.1. Introduction 223
   9.2. Materials and methods 225
       9.2.1. Sample material 225
       9.2.2. Experimental approach 225
       9.2.3. Kinetic of acid ester hydrolysis 226
   9.3. Results and discussion 229
       9.3.1. Reaction rate constants of the acid hydrolysis to cleave formate and acetate from coal samples 229
       9.3.2. Critical evaluation of the determination of k' 237
       9.3.3. Temperature dependence of the reaction rate constant for the acid hydrolysis of formate and acetate 239
       9.3.4. Problems in simulating the natural hydrolysis process in laboratory scale 243
   9.4. Conclusion 245

10. **Ether Cleavage In New Zealand Coal Samples**   **247**
   10.1 Introduction 247
   10.2 Materials and Methods 248
       10.2.1 Sample material 248
       10.2.2 Methods 249
   10.3 Results and discussion 251
       10.3.1 Qualitative evaluation of the BBr_3 cleavage products from the New Zealand coal series 251
       10.3.2 Quantitative evaluation of the BBr_3 cleavage products from the New Zealand coals 259
   10.4 Conclusion 263

11. **Conclusions**   **265**

References   **310**

Appendix   **311**

A. Tables	313
B. Mass spectra	325
C. Publications	331
Acknowledgements	333

List of Figures

2.1. The tree of life as defined by comparative rRNA gene sequencing showing the three domains of life *Bacteria, Archaea* and *Eukarya*, modified from Madigan and Martinko (2006). 40

2.2. Bacterial distributions in the Juan de Fuca hydrothermal vent field (IODP Leg 169), Bent Hill Massive Sulphide site, with respect to depth and temperature, taken from Parkes et al. (2000). 43

2.3. The deep hydrogen-driven biosphere hypothesis, illustrated by its carbon cycle. 47

2.4. The photosythetic production by phytoplancton decreases with decreasing light intensity but respiration appears throughout the whole water column. At the compensation depth phytoplanctonic production and respiration intensity is equal. The critical depth is reached when all phytoplanktonic matter is respirated. Taken from Killops and Killops (2004) 52

2.5. Evolution pathway of the three different kerogen types (a) and coal macerals (b) upon geological burial in a van Krevelen diagram of H/C as O/C ratios. 54

2.6. Distribution of oxygen containing functional groups in the three different types of kerogen, adapted from Killops and Killops (2004). 56

2.7. Chemical structures of the three kerogen types at the beginning of diagenesis, after Béhar and Vandenbroucke (1987). 57

2.8. Scheme of kerogen evolution showing diagenesis, catagenesis and metagenesis stage, including oil and gas generation zones, modified after Tissot and Welte (1984b). 60

2.9. Theoretical structures of a Type III kerogen at different maturity stages, from Béhar and Vandenbroucke (1987). 64

2.10 World coal consumption in million short tons and world CO_2 emission in metric tons. 66

2.11 Random, partial structure of lignin after Nimz (1974) with colored main lignin building blocks coumaryl, coniferyl and sinapyl alcohol. 69

2.12 Stages of organic matter evolution and comparisation of different maturity scales for coals and kerogen, modified after Tissot and Welte (1984b) and Vandenbroucke and Largeau (2007) . . . 72

2.13 Maturation stages of coals plotted in a van Krevelen type diagram. 73

2.14 Reactions transformig lignin to lignite, sub-bituminous coal and high volatile bituminous coal, modified after Hatcher and Clifford (1997) and Killops and Killops (2004). 74

2.15 The electromagnetic spectrum from low energy (long wavelength, AM) to high energy content (short wavelength, gamma ray) can be used for analytical techniques. 77

2.16 Franck-Condon principle 79

2.17 The maceral liptinite in fluorescence microscopy. 80

2.18 IR spectra of kerogen of different maturation. 81

2.19 Separation of electron spin moment by an external strong magnetic field (B) into two spin states (+1/2 and -1/2) with the energy difference ΔE (Zeeman effect, g: g-factor, μ_B: bohr magneton). 83

2.20 Scheme of Rock-Eval pyrolysis, showing the different fractions of total organic matter in a rock sample, taken from Lafargue et al. (1998). 86

2.21 Scheme of a stepwise chemical degradation procedure. 89

2.22 Scheme of ester cleavage in coals, liberating terminally bound organic acids and alcohols. The catalysts for this reaction can be either protons (acid hydrolysis) or hydroxy ions (alkaline hydrolysis, saponification). 90

2.23 Scheme of ether cleavage in coals, liberating terminal bound alcohols. 91

2.24 Scheme of C-S and S-S cleavage in coals, liberating terminal sulphur bound compounds. 92

2.25 Scheme of the oxidation of coal with RuO_4 resulting in aliphatic and aromatic carboxylic acids. 93

3.1. Map of New Zealand, showing the location of the samples chosen for this study, the sedimentary basins, the coalfields, the three basement zones: Eastern Province, Western Province and Median Tectonic Zone (dotted line: range of the basement zones) and the location of the DEBITS-1 well. 96

3.2. Stratigraphy of the DEBITS-1 well including formations, age and stratigraphic groups. 101

4.1. Flow chart showing the total sample processing applied to the lignite and coal samples in this thesis. 112

5.1. Experimental procedure for analysis of LMWOAs released from coals by alkaline ester cleavage reaction. 127

5.2. Recovery of formate and acetate after ester cleavage of its ethylated congeners using different analytical ester cleavage conditions. 131

5.3. Formate, acetate and oxalate concentrations liberated by ester cleavage reaction from New Zealand coals of different maturity. 134

6.1. Separation scheme outlining the analysis of ester-bound fatty acids and alcohols from coal samples using GC-MS and open system pyrolysis-GC. 151

6.2. GC-MS-chromatograms of alcohol and fatty acid fractions of a low-rank (G001978) and of a moderate-rank (G001993) sample. 154

6.3. Concentrations (in mg/gTOC) of total alcohols (a) and total fatty acids (b) released by alkaline ester cleavage from a series of New Zealand coal samples of different maturity. 156

6.4. Alcohol (a-e) and fatty acid (FA; f-k) distribution patterns of selected coal samples from early to moderate coalification level (0.27% to 0.80% R_0). 157

6.5. Concentrations (in mg/g TOC) of long chain fatty acids (a) and short chain fatty acids (b) released by alkaline ester cleavage from a series of New Zealand coal samples of different maturity. Carbon preference index (CPI_{FA}) for long chain fatty acids (C_{20}-C_{30}) (c) and short chain fatty acids (C_{14}-C_{18}) (d) versus sample maturity. 159

6.6. Distribution pattern of alkenes obtained by open system pyrolysis of sample G001978 (Eastern Southland Basin) from early diagenesis (R_0: 0.28%) (a) and of sample G001982 (Waikato Basin) from late diagenesis (R_0: 0.40%) (b) before and after alkaline ester cleavage. For comparison, the corresponding distribution patterns of ester-bound alcohols (c and d) and fatty acids (e and f) obtained after alkaline cleavage reaction are shown. 167

6.7. Distribution of ω-hydroxy fatty acids detected in the Northland and Eastern Southland Basin samples after saponification. . . . 171

6.8. Distribution of α-ω-dicarboxylic acids found in the Northland Basin samples after saponification. 172

6.9. Structures of the pentacyclic triterpenoid alcohols betulin and lupeol. 174

7.1. Stepwise procedure for the fractionation of bitumen for the analysis of free fatty acids, alcohols and esters. MPLC: medium pressure liquid chromatography (Radke et al., 1980), NSO: nitrogen, sulfur and oxygen fraction. 179

7.2. Distribution of free fatty acids in the bitumen of the immature sample G001976 from the Eastern Southland Basin (m/z=74) . 181

7.3. Concentration trend of the total free fatty acids in the bitumen fraction of the NZ coal samples of different maturity and additionally in total and separated into short chain (n-C_{14} to n-C_{19}) and long chain fatty acids (n-C_{20} to n-C_{32}). 182

7.4. Calculated carbon preference index for fatty acids (CPI_{FA}) separated for short chain and long chain fatty acids (FAs) vs. maturity. 183

7.5. Concentrations of the total alcohols in the bitumen fraction of the NZ coal samples of different maturity. 186

7.6. Distribution of the free alcohols in the bitumen of the immature samples from the DEBITS-1 well (Waikato Basin) and from the Eastern Southland Basin. 187

7.7. Distribution of the fatty acid methyl esters (FAME) and the calculated CPI_{FA} in the bitumen of NZ coal samples, separated for long chain and short chain esters. 188

7.8. Mass spectra of the unknown pentacyclic triterpenoids U1, U2 and U3 detected in the neutral fraction of the bitumen extract from the NZ coal samples. 191

7.9. Structures of identified pentacyclic triterpenoids. 192

7.10 Percentage proportions in % of short chain fatty acids ($C_{13/14}$ to C_{19}) and long chain fatty acids (C_{20} to $C_{30/32}$) in the kerogen-bound fatty acid fraction (A) and the free fatty acid fraction (B). 195

7.11 The proportion of the kerogen fatty acids (KFA) to the total fatty acids (kerogen fatty acids (KFA) + bitumen fatty acids (BFA) indicating the different susceptibility of free and kerogen-bound fatty acids to diagenetic processes. 198

8.1. Ester-bound LMWOAs from lignite and sub-bituminous coal layers of the DEBITS-1 well. 209

8.2. Ester-bound total as well as short chain (C_{13}-C_{19}) and long chain (C_{20}-C_{32}) fatty acids and total alcohols from lignite and coal layers of the DEBITS-1 well and carbon preference index data of short chain and long chain fatty acids and alcohols. . . . 212

8.3. Ester-bound LMWOA from a sandstone, silt, clay and lignite transect in subcore 9.2 (25.93 - 26.15 m), compared with glycerol dialkyl glycerol tetraethers (GDGT), total phospholipids (TPL), porosity, permeability and TOC. 215

9.1. Experimental procedure to investigate the kinetics of acid-catalysed ester hydrolysis of coal samples. 225

9.2. Residual kerogen-bound formic (A) and acetic acids (B) in the NZ coal samples during 7 d of hydrolysis and determination of k' for hydrolysis of formic (C) and acetic acid esters (D) (90°C, pH 3) by linear regression. 230

9.3. Reaction rate constants k' for hydrolysis of formic (A) and acetic esters (B) from the kerogen matrix of NZ coal samples of different maturity at 90°C and pH 3. 230

9.4. Residual kerogen-bound formic (A) and acetic acids (B) in the NZ coal samples during 7 d of hydrolysis and determination of k' for hydrolysis of formic (C) and acetic acid esters (D) (75°C, pH 3) by linear regression. 232

9.5. Residual kerogen-bound formic (A) and acetic acids (B) in the NZ coal samples during 7 d of hydrolysis and determination of k' for hydrolysis of formic (C) and acetic acid esters (D) (75°C, pH 3) by linear regression. 232

9.6. Residual kerogen-bound formic (A) and acetic acids (B) in the NZ coal samples during 7 d of hydrolysis and determination of k' for hydrolysis of formic (C) and acetic acid esters (D) (60°C, pH 3) by linear regression. 234

9.7. Residual kerogen-bound formic (A) and acetic acids (B) in the NZ coal samples during 7 d of hydrolysis and determination of k' for hydrolysis of formic (C) and acetic acid esters (D) (60°C, pH 3) by linear regression. 236

9.8. Residual kerogen-bound formic (A) and acetic acids (B) in the NZ coal samples during 7 d of hydrolysis and determination of k' for hydrolysis of formic (C) and acetic acid esters (D) (45°C, pH 3) by linear regression. 236

9.9. Arrhenius plot for the acidic hydrolysis of A) formic acid ester and B,C,D) acetic acid ester in NZ coal samples and linear regression to calculate activation energy and frequency factor. . . 242

10.1.(A) Total ion current (TIC) chromatogram of a monobromoalkane standard mix (n-C_{12}, n-C_{14}, n-C_{16}, n-C_{18}) and (B) TIC-chromatogram of BBr_3 cleavage products from sample G001978 (Eastern Southland Basin, R_0 0.28%), (C) all brominated compounds with more than three carbon atoms (mass trace m/z 135), (D) ethyl esters (mass trace m/z 88). 251

10.2 Selected structures of detected aromatics, phenols, ethyl esters and brominated esters and ketones from the New Zealand coal samples forming sub-structures within the macromolecular organic matter. The position of bromine reveals the former ether linkage, the ethyl ester is suggested to derive from former ester linkage. 257

10.3 Products of BBr_3 ether cleavage of NZ lignite and coal samples of different maturity, grouped into mono-brominated alkanes (A), poly-brominated alkanes (B), ethyl esters (C) as well as brominated esters and ketones (D). Note different y-axis. 260

B.1. MS of aliphatic monobro-malkanes 325
B.2. MS of tribromo-propane 326
B.3. MS of dibromo-butane 326

B.4. MS of tribromo-pentane 327
B.5. MS of tetrabromo.pentane 327
B.6. MS of dibromo-propan-2-one 328
B.7. MS of dibromo-trimethyl-benzene 328
B.8. MS of bromo-acetc-acid-ethyl-ester 329
B.9. MS of 3-bromo-butanole-acetate 329

List of Tables

2.1. Redox reactions providing energy (G) for microorganisms, modified after McKinley (2001). 46

2.2. Petrographic classification of bituminous coal constituents (Bend, 1992) including maceral groups and related macerals and the originating material. 68

3.1. Selected samples from different coal mines of the North and the South Island of New Zealand, including (GFZ) sample number, sample location, age and maturity rank data, provided by Richard Sykes, GNS Science (New Zealand). S_R=Suggate rank (Suggate, 2000). 97

3.2. Geochemical data for the coal sample series selected for the current study, provided by R. Sykes, GNS Science. 98

3.3. Selected samples from DEBITS-1 core, including depth [m below surface], stratigraphic groups, subcore number, lithology, TOC content and vitrinite reflectance data. TOC content and R_0 data were provided by R. Sykes (GNS Science, NZ). 102

5.1. Selected samples for the investigation of kerogen bound formate, acetate and oxalate including information about basin, coalfield, formation, age, vitrinite reflectance and TOC (data provided by R. Sykes, GNS Science (NZ)) and the detected amounts of formate, acetate and oxalate. 125

5.2. Assessment of required formate, acetate and oxalate respiration rates to sustain deep microbial life (modified after (Horsfield et al., 2006)). 137

5.3. Calculated feedstock potential of the New Zealand coals investigated. 139

6.1. Information about the sample maturities (vitrinite reflectance R_0), sedimentary basins, lithological formations, total organic carbon (TOC) contents, ester-bound fatty acid and alcohol contents and carbon preference indices (CPI_{FA}) of short (C_{14}-C_{18}) and long (C_{20}-C_{30}) chain fatty acids of the New Zealand coal samples. 148

6.2. Concentrations of dicarboxylic acids, ω-hydroxy acids and functionalised pentacyclic triterpenoids detected in the NZ lignite samples. 173

7.1. Total amounts of free fatty acids (FA) from the bitumen fraction (in mg/gTOC) and calculated CPI_{FA} values separated for short chain and long chain fatty acids. 184

7.2. Total amounts of free alcohols (ALC) and fatty acid methyl esters (FAME) from the bitumen fraction (in μg/gTOC) and calculated CPI_{FAME} values separated for short chain and long chain fatty acids. 185

7.3. Pentacyclic triterpenoids detected in the Eastern Southland Basin and three Waikato Basin samples. 189

7.4. Percentage proportions of short chain fatty acids (SCFA, $C_{12/13}$ to C_{19}) and long chain fatty acids (LCFA, C_{20} to $C_{30/32}$) of the amount of total fatty acids in the kerogen-bound and the free (bitumen) fatty acid fraction. 193

7.5. Amounts of all kerogen bound fatty acids (KFA) and bitumen fatty acids (BFA) and total amount of fatty acids (KFA + BFA), percentage proportions of KFA and BFA of the total fatty acids and the ratio of KFA to the sum of KFA + BFA indicating the dominance of bound or free fatty acids. 197

8.1. LMWOAs released from lignite and sub-bituminous samples of DEBITS-1 well subcores 6.3, 9.2, 11.1, 17.4, 18.2, 29.5 and 32.2 by alkaline ester cleavage reaction in mg/gTOC. 208

8.2. Fatty acids and alcohols released from organic matter rich layers of DEBIRT-1 well and calculated CPI for released fatty acids and alcohols. 211

8.3. Low molecular weight organic acids released from samples of a sand to lignite transsect of the DEBITS-1 well of 9.2 obtained after alkaline ester cleavage reaction in mg/gSed. Additionally listed are the data for the glycerol dialkyl glycerol tetraethers (GDGTs), total phospholipids (TPL), porosity and permeability, provided by K. Mangelsdorf (GFZ Potsdam) 217

9.1. Determined reaction constant k' (pH 3) from the slope of the linear regression in figure 9.2, calculated k' (pH 7) using equation 9.8 and half life ($t_{1/2}$) using equation 9.13 of the ester concentration for the hydrolysis of formic and acetic acid ester in NZ coal samples at 90°C. 231

9.2. Determined reaction constant k' (pH 3) from the slope of the linear regression in figure 9.2, calculated k' (pH 7) using equation 9.8 and half life ($t_{1/2}$) using equation 9.13 of the ester concentration for the hydrolysis of formic and acetic acid ester in NZ coal samples at 75°C. 233

9.3. Determined reaction constant k' (pH 3) from the slope of the linear regression in figure 9.2, calculated k' (pH 7) using equation 9.8 and half life ($t_{1/2}$) using equation 9.13 of the ester concentration for the hydrolysis of formic and acetic acid ester in NZ coal samples at 60°C. 235

9.4. Determined reaction constant k' (pH 3) from the slope of the linear regression in figure 9.2, calculated k' (pH 7) using equation 9.8 and half life ($t_{1/2}$) using equation 9.13 of the ester concentration for the hydrolysis of formic and acetic acid ester in NZ coal samples at 45°C. 237

9.5. Reaction rate constant k' values determined during the kinetic experiments. 241

9.6. Calculation of E_A and A from linear regression (Figure 9.9) for the hydrolysis of formic and acetic acid ester in NZ coal samples. 243

10.1 Monobromo-, dibromo-, tribromo- and tetrabromo-alkanes after BBr_3 ether cleavage experiments from NZ coal samples ranging from diagenesis to the main catagenesis stage. 254

10.2 Brominated esters and ketones, aromates, phenols and ethyl esters after BBr_3 ether cleavage experiments from NZ coal samples ranging from diagenesis to the main catagenesis stage. . . . 255

A.1. Hydrolysis of formic and acetic acid esters (90°C, pH 3) for 1, 2, 3, 4 and 7 d of sample G004541. 314

A.2. Hydrolysis of formic and acetic acid esters (90°C, pH 3) for 1, 2, 3, 4 and 7 d of sample G001980. 314

A.3. Hydrolysis of formic and acetic acid esters (90°C, pH 3) for 1, 2, 3, 4 and 7 d of sample G001996. 315

A.4. Hydrolysis of formic and acetic acid esters (90°C, pH 3) for 1, 2, 3, 4 and 7 d of sample G001989. 315

A.5. Hydrolysis of formic and acetic acid esters (75°C, pH 3) for 1, 2, 3, 4 and 7 d of sample G004541. 316

A.6. Hydrolysis of formic and acetic acid esters (75°C, pH 3) for 1, 2, 3, 4 and 7 d of sample G001980. 316

A.7. Hydrolysis of formic and acetic acid esters (75°C, pH 3) for 1, 2, 3, 4 and 7 d of sample G001996. 317

A.8. Hydrolysis of formic and acetic acid esters (75°C, pH 3) for 1, 2, 3, 4 and 7 d of sample G001989. 317

A.9. Hydrolysis of formic and acetic acid esters (60°C, pH 3) for 1, 2, 3, 4 and 7 d of sample G004541. 318

A.10 Hydrolysis of formic and acetic acid esters (60°C, pH 3) for 1, 2, 3, 4 and 7 d of sample G001980. 318

A.11 Hydrolysis of formic and acetic acid esters (60°C, pH 3) for 1, 2, 3, 4 and 7 d of sample G001996. 319

A.12 Hydrolysis of formic and acetic acid esters (60°C, pH 3) for 1, 2, 3, 4 and 7 d of sample G001989. 319

A.13 Hydrolysis of formic and acetic acid esters (60°C, pH 3) for 1, 2, 3, 4 and 7 d of sample G004541. 320

A.14 Hydrolysis of formic and acetic acid esters (45°C, pH 3) for 1, 2, 3, 4 and 7 d of sample G001980. 320

A.15 Hydrolysis of formic and acetic acid esters (60°C, pH 3) for 1, 2, 3, 4 and 7 d of sample G001996. 321

A.16 Hydrolysis of formic and acetic acid esters (45°C, pH 3) for 1, 2, 3, 4 and 7 d of sample G001989. 321

A.17 Data calculation for *Arrhenius* plot of sample G004541. 322

A.18 Data calculation for *Arrhenius* plot of sample G001980. 322

A.19 Data calculation for *Arrhenius* plot of sample G001996. 323

A.20 Data calculation for *Arrhenius* plot of sample G001989. 323

Abbreviations

ALC	alcohol
BFA	bitumen fatty acids
BI	bitumen index
bit.	bituminous
CC	column chromatography
CP	cross polarisation
CPI	carbon preference index
CT	computed tomography
daf	dry ash free
dmmsf	dry, mineral matter (and) sulfur free
DEBITS	Deep Biosphere In Terrestrial Systems (Project of GFZ Potsdam (GER), GNS Science (NZ) and Cardiff University (UK))
DOE	Department of Energy
EIA	Energy Information Administration
EPR	electron paramagnetic resonance
ESB	Eastern Southland Basin
ESI	electron spray ionisation
ESR	electron spin resonance
EtOH	ethanol
GC	gas chromatography
FA	fatty acid
FAEE	fatty acid ethyl ester
FAME	fatty acid methyl ester

FID	flame ionisation detector
FIR	far-infrared
FTIR	fourier transformation infrared spectroscopy
GRS	Green River Shale
HI	hydrogen index
HC	hydrocarbon
HI	hydrogen index
HMW	high molecular weight
HPLC	high performance liquid chromatography
IC	ion chromatography
ICCP	International Commission for Coal Petrology
ICDP	International Continental Scientiffic Drilling Program
IODP	Integrated Ocean Drilling Program
ISTD	internal standard
IR	infrared
KFA	kerogen (bound) fatty acids
LMW	low molecular weight
LMWOA	low molecular weight organic acids
MAS	magic spining angel
MeOH	methanol
MIR	mid-infrared
mmf	mineral matter free
MPLC	medium pressure liquid chromatography
MS	mass spectrometry
NB	Northland Basin
NIR	near-infrared
NMR	nuclear magnetic resonance
NZ	New Zealand
OI	oxygen index
OM	organic matter
PI	production index

QI	quality index
R_0	vitrinite reflectance [%]
ROC	reduced [form of] organic carbon
r. t.	room temperature (approx. 20°C)
S_R	Suggate rank
TB	Taranaki Basin
TMSCl	trimethyl silyl chloride
TMSI	trimethyl silyl iodide
TOC	total organic carbon
UV	ultraviolet
Vis	visible
vol.	volatile
VR	vitrinite reflectance
WB	Waikato Basin
WCB	West Coast Basin
XANES	X-ray absorbtion near-edge structure

1. Introduction

In recent years, it has been recognized that microorganisms occur widely disseminated in the deep subsurface of the Earth, far away from any direct photosynthetic energy supply (Fredrickson and Onstott, 1996; Parkes et al., 2000; Pedersen, 2000). It is assessed that the biomass of this so-called deep biosphere at least equals that of the surface biosphere (Whiteman et al., 1998). Therefore, the deep biosphere must play a fundamental role in global biogeochemical cycles over both, short and long time scales. It has been stated that diverse microbial life appears to be present wherever a source of energy (substrate) is available (Kerr, 1997). Main limiting factors appear to be in addition to substrate and nutrient supply, temperature and available pore space inside the rocks with sufficient permeability. Due to their widespread occurrence, microorganisms contribute to all elemental and material cycles, however, their individual role and their implication is still far from being completely understood.

With the discovery of this deep subsurface microbial life, the question on its potential carbon and energy sources arises. Depending on the habitat conditions, the subsurface microbial communities are able to utilize various sources and mechanisms to gain their energy and substrates. In igneous rocks (e.g. basalts and granites), lithoautotrophic microbial communities are able to synthesis small organic compounds from inorganic sources like radiolytically generated hydrogen gas and inorganic carbon dioxide, and these simple organic compounds can be utilized by other microorganisms. In marine and terrestrial environments, the most obvious carbon and energy source is buried organic matter. The hypothesis investigated in this study

1. Introduction

is that organic carbon rich layers, like coal deposits, can provide a significant carbon and energy source for deep microbial populations (*Bacteria* and *Archaea*). During maturation, this sedimentary organic matter undergoes several biotic and abiotic alteration processes and thereby loses functional groups containing oxygen, partly in form of CO_2, low molecular weight organic acids (e.g. acetic acid) or fatty acids. Therefore, the organic matter rich layers (*feeder lithologies*) may be able to feed microbial communities inhabiting adjacent sedimentary layers (*carrier lithologies*). Horsfield et al. (2006) were able to show an increase of microbial activity in a deep sedimentary interval with increased thermally induced substrate release, which points to a coupling of biological and geological processes (*bio-geo coupling*).

This thesis is part of the DEBITS (Deep Biosphere In Terrestrial Systems) project, commenced in February 2004 as a joint interdisciplinary collaboration of geologists and geochemists from *GFZ Potsdam*[1], *GNS Science*[2] and microbiologists from the *University of Cardiff*[3]. The DEBITS project is exclusively dedicated to the investigation of deep terrestrial microbial ecosystems. The study area of this project is New Zealand.

New Zealand represents a perfect natural laboratory to investigate the bio-geo coupling hypothesis. The so-called New Zealand coal band contains a coal series of almost continuous maturity from Cretaceous to Pleistocene age that represents, therefore, different feeding potentials. Within the scope of the DEBITS project, coal samples of different maturity ranging from lignite to sub-bituminous coal were gathered from various coal mines and outcrops on the North and South Island of New Zealand. Additionally, organic carbon rich samples were taken from the 148 m deep DEBITS-1 well, which was drilled in the Waikato coal area in February/March 2004. The DEBITS core covers a sedimentary section of sandstones, siltstones, claystones and

---

[1]Helmholtz Centre Potsdam, German Research Centre for Geosciences (GFZ), Telegrafenberg, 14473 Potsdam, Germany

[2]Institute of Geological & Nuclear Sciences, PO Box 30368, Lower Hutt 5040, New Zealand

[3]School of Earth, Ocean and Planetary Sciences, Cardiff University, Wales, Cardiff CF10 3XQ, United Kingdom

imbedded lignite and coal layers. Therefore, it provides a suitable sample material to investigate the feeder-carrier hypothesis.

Within the scope of the DEBITS project in this thesis, potential substrates and habitats for deep microbial life associated with structural alteration of buried organic matter are investigated by addressing the following questions:

- How does the oxygen containing compound fraction (esters and ethers) bound to the kerogen matrix change with increasing maturation of the organic matter and do organic carbon rich lithologies such as lignites and coals provide a significant feedstock for deep microbial life in terrestrial settings e.g. by a constant release of low molecular weight organic acids such as formate and acetate during ongoing maturation processes?

- Are there indications for microbial communities being stimulated by the geothermally induced release of hydrocarbons during the early catagenesis when temperature conditions are still compatible with microbial life as outlined in the hypothesis of bio-geo coupling?

- Are organic matter rich lithologies able to feed deep microbial life in adjacent organic carbon poor lithologies by releasing important substrates for microbial metabolism into the surrounding (feeder-carrier hypothesis)?

- Is aqueous hydrolysis in the pore water within lignites and coals an appropriate process for the release of potential substrates from the organic matter during maturation and does the structural alteration during maturation affect this hydrolysis process and, therefore, the release of the substrates?

- What is the role of oxygen containing compounds in forming the macromolecular network structure and how does this compound fraction change with the structural alterations during ongoing diagenetic and catagenetic maturation?

1. Introduction

This thesis is structured as follows. The following Chapter (2) elucidates the theoretical background of this study. This chapter is a literature review on the basic background information and recent developments. It focuses mainly on three topics: [1] the discovery of a deep biosphere and its contribution to elemental cycles within the geosphere (Section 2.1), [2] changes in the chemical structure during the formation, alteration and maturation of coal and kerogen (Section 2.2) and [3] important methods for revealing structural information from kerogen, including instrumental analytical methods and chemical degradation reactions (Section 2.3).

Chapter 3 describes the study area New Zealand and its geological settings and the methods applied in this study are outlined in Chapter 4.

The development of an analytical approach for analysis of low molecular weight organic acids chemically bound to coals as well as the results of its application to selected samples taken from the New Zealand coal band to estimate the release of potential substrates during maturation is the content of Chapter 5. The distribution of high molecular weight fatty acids and alcohols that are chemically linked via ester bonds to the coal samples is discussed in Chapter 6. In Chapter 7, the distribution of free fatty acids and alcohols from the bitumen fraction of these coals is investigated and compared with the corresponding kerogen-bound compounds reported in Chapter 6.

In Chapter 8, the amounts of ester bound low molecular weight organic acids as well as fatty acids and alcohols from organic carbon rich layers of the DEBITS-1 well related to burial depth are investigated. A 25 cm long transect from a lignite layer into adjacent silt- and sandstone layers is selected in order to examine the substrate potential of these lithologies and whether microbial communities are associated to this organic carbon rich lithology.

In Chapter 9, the kinetics of proton catalysed hydrolysis of ester-bound compounds suggested to be an appropriate process for releasing potential substrates from lignites and coals is presented. Finally, Chapter 10 describes the cleavage of alcohols that are bound via one ore more ether bonds to the NZ lignites and coals in order to obtain a deeper insight into structural properties within the kerogen matrix.

2. Background

2.1. Microorganisms and the Deep Biosphere

2.1.1. The world of microbes

The term *microorganism* (also named microbes) is deduced from the greek words *micrós* (small) and *organismós* (organism). It is used for microscopic organisms that are usually too small to be visible by the naked human eye. They were first discovered in 1675 by the dutch scientist Antonie Philips van Leeuwenhoek (1632-1723) by using an improved microscope and who was, therefore, considered as the first microbiologist.

The world of microorganisms contains various forms of small organisms, such as bacteria, archaea, fungi, protists (such as green algae), animals (such as zooplankton), planarian and amoebae. Microorganisms are part of all three domains of life, the *Bacteria*, the *Archaea* and the *Eukarya* (Figure 2.1) (Madigan and Martinko, 2006). *Bacteria* and *Archaea* are prokaryotes, they lack of a cell nucleus and other membrane bound organelles. Eukaryotes (including protists, fungi, animals and plants) contain cell organelles such as the cell nucleus. Most of the microorganisms are single-celled organisms but this is not universal because of the existence of some multicellular microscopic organisms and some rare single-celled macroscopic bacteria that are visible by the naked human eye.

Up to now microbes have been found in various habitats both above and beneath the Earth's surface. They provide essential services to other

2. Background

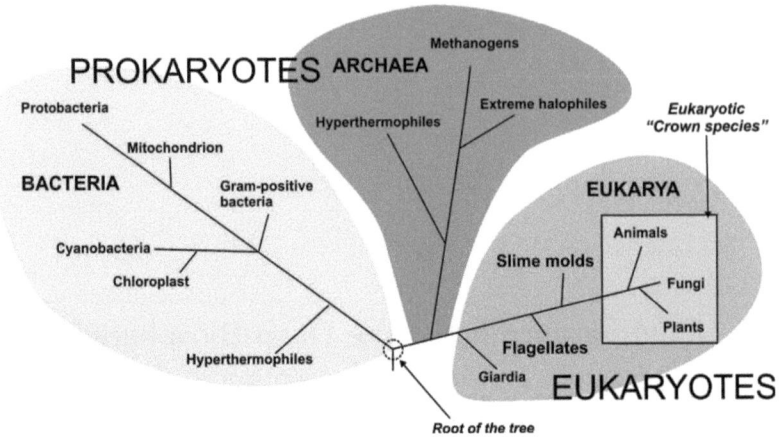

Figure 2.1.: *The tree of life as defined by comparative rRNA gene sequencing showing the three domains of life Bacteria, Archaea and Eukarya, modified from Madigan and Martinko (2006).*

creatures by composting waste and forming nutrients. Some of them are associated in symbiosis with other life forms. For example, hundreds of species and billions of individual microorganisms grow in or on the human body (referred to as normal microbial flora) and some of them contribute directly to our health (Madigan and Martinko, 2006). However, some of them called parasites or often pathogens can cause damage to the host and cause diseases.

Microorganisms can reproduce rapidly and some (such as bacteria) are able to freely exchange genes with other species. The high mutation rate allows a rapid evolution by natural selection and helps the microbes to adapt very fast to different environmental conditions. Therefore, microbes can be found in various and sometimes extreme environments, such as the deep sea, polar- and permafrost areas, hot surface springs and deep sea hydrothermal vents, hypersaline and deep water lakes, mines and oil reservoirs and the deep subsurface of Earth. They are the only life form expected to inhabit the sediments of extraterrestrial planets such as mars (Rothschild,

2.1.2. Discovering the Deep Biosphere

It was in the nineteen twenties that the first hints were found indicating microbial life may appear hundreds to thousands of meters below the surface. Edson Sunderland Bastin (1878-1953), a geologist from the University of Chicago, wondered why water extracts from oil fields contained hydrogen sulfide and bicarbonate. He speculated that so-called sulfate reducing bacteria (SRB) must live in the subsurface oil reservoirs and produce hydrogen sulfide and bicarbonate by degrading organic compounds in the oil. Together with his colleague Frank E. Greer, he cultivated sulfate reducing bacteria from groundwater samples extracted from an oil reservoir hundreds of meters below the surface and speculated that the detected microbes are posterity of former organisms buried 300 million years ago when these sediments were deposited (Fredrickson and Onstott, 1996). But being aware that oil drilling techniques were not designed to drill uncontaminated samples, they could not exclude that the microorganisms they observed were introduced from the surface. Thus, they achieved only little support for their theory in the scientific community. Therefore, to ensure that drilled samples are free of any contaminant from the surface is one of the most crucial and challenging tasks when exploring the deep biosphere, even to this day.

Until the mid of the 20th century, the scientific dogma held, that the deeper subsurface was sterile. In the nineteen fifties, Claude Ephraim ZoBell (1904-1989) and his colleagues of the Scripps Institution of Oceanography investigated microbial processes within the sediments of the seabed. Due to their inability to culture bacteria at greater depth, they defined that the base of marine biosphere is found at 7.47 m of depth (Morita and ZoBell, 1955).

During the late 70th and early 80th, the U.S. Geological Survey and the Environmental Protection Agency reevaluated their understanding of

2. Background

groundwater chemistry including considerations of possible microbial inhabitants within water-yielding rock formations. At this time, the U.S. Department of Energy (DOE) was searching for a possibility of cleaning up the soil of industrial facilities where they dumped vast amounts of waste (organic rich solutions, metals and radioactive materials) during the cold war. The DOE geologist Frank J. Wobber argued that if microbes are present below the earth's surface, they could help to degrade buried organic pollutants. He initiated the DOE's Subsurface Science Program to address a systematically research on deep buried microbial communities with the help of biologists, geologists and chemists (Fredrickson and Onstott, 1996). In the scope of this research program, new techniques of contamination free drilling and sampling were developed.

From that time on, the research on deep biosphere has extended intensively and diverse microbial communities in different habitats were found within the following decades. In 1992 drillers of Texaco and the Eastern Exploration Co. found a repository of life at 2.8 km of depth when they were searching for oil in a 4 km deep rift (Taylorsville Basin, USA) (Kerr, 1997). In 1994, R. John Parkes of the University of Bristol (UK) investigated microbial communities in sediments of the Pacific Ocean from beneath more than 500 m depth (Parkes et al., 1994). In 1997, Tullis C. Onstott reported the finding of sulfate- and iron-reducing bacteria in gold mines in the Witwatersrand Supergroup in South Africa. These colonies were buried in ca. 3500 m of depth at an ambient temperature of 50°C (Onstott et al., 1997).

Within the scope of IODP Leg 169 focussed on the Juan de Fuca hydrothermal vent field Parkes et al. (2000) reported that so-called mesophilic bacteria were found up to a temperature limit of 45°C. Deeper buried bacteria are thermophilic bacteria, which can grow up to 80°C. Deeper still they have to be hyperthermophiles that can grow at temperatures higher than 80°C. (Figure 2.2).

The current established upper temperature limit for bacterial grows (113°C) was published by Blöchl et al. (1997). Higher temperatures up to

2.1. Microorganisms and the Deep Biosphere

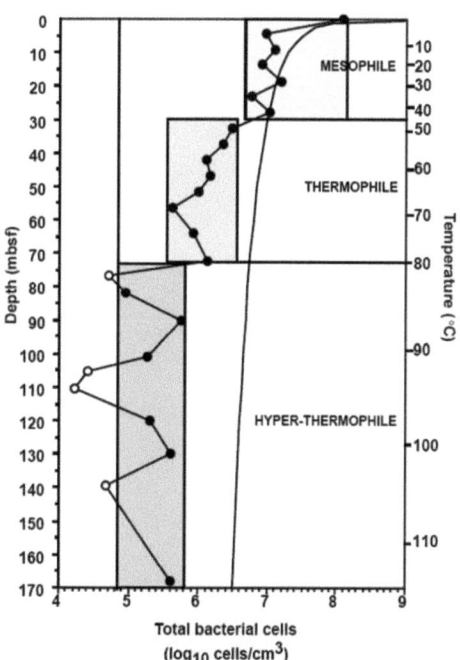

Figure 2.2.: *Bacterial distributions in the Juan de Fuca hydrothermal vent field (IODP Leg 169), Bent Hill Massive Sulphide site, with respect to depth and temperature, taken from Parkes et al. (2000).*

121°C (Kashefi and Lovley, 2003) are still discussed controversially in the literature (Cowan, 2004; Kashefi, 2004). After some decades of research, it seems that life is present wherever a source of energy is present and the only true limit on depth appears to be temperature (Kerr, 1997).

2.1.3. Everything is *not* everywhere

In 1934, the dutch microbiologist L.M.G. Bass-Becking formulated a statement that has become a fundamental paradigm in microbial ecology for nearly a century (Cho and Tiedje, 2000). His famous sentence "Everything is

43

2. Background

everywhere, the environment selects" (Baas-Becking et al., 1934) represents the opinion that the same types of bacteria can be present everywhere. To support his theory, he argued that bacteria are dispersed from one area of the world to another by wind and water as well as by birds that fly between the continents and carry bacteria with them.

Recent developments on DNA/RNA analysis have helped to investigate this hypothesis. It turned out that some types of bacteria are indeed cosmopolitan (e.g. partial sequences of 16S rDNA of two strains of *Microcoleus chthonoplastes* from Europe and two from North America are identical), but in other cases, bacterial colonies were found to be independent (Staley, 1999). For example, Castenholz (1996) found out that some thermophilic cyanobacteria were not found in North American hot springs, but were found in those of Alaska and Iceland, inconsistent with the cosmopolitan hypothesis.

With ongoing research on microbial genomes, it becomes more and more obvious that microbial life is of an enormous complexity regarding the number of different species of microbes and with each new study, the window on the microbial world increases in size by the discovery of previously unknown microorganisms (Sogin et al., 2006). For example, Roussel et al. (2008) reported the finding of the up to now deepest evidence for prokaryotic cells of the sub-seafloor biosphere at 1600 mbsf in sediment samples from the Newfoundland Margin, being about 111 Ma old and living at temperatures of 60° to 100°C.

Temperature is a crucial factor for setting the boundaries of life in the subsurface, however, there are other factors influencing the distribution and the activity of microorganisms in the sediments as well. Past and present geological processes thereby dictate the conditions of the microbial life habitats. Relevant factors are the rate and age of water flowing through the system as well as the availability of carbon and energy sources, being influenced by the geological history and the chemical composition of the rock or sediment. In this context, porosity and permeability of the rock or sediment are also of major importance (Fredrickson and Onstott, 2001).

2.1.4. Feeding of the Deep Biosphere - driving Earth's biogeochemical cycles

As any living organism microorganisms need a source of carbon and energy to survive. They are experts on utilizing any energy that becomes thermodynamically available. This energy can be extracted from various sources, depending on the geological environment. In the subsurface, where sunlight is absent and no photosynthesis is possible, the microbes gain their energy from chemical (redox) reactions. The organisms thereby act as redox catalysts mediating the electron transfer reaction including a substrate oxidizing half-reaction and a coupled electron acceptor half-reaction. The available free energy (ΔG) is used by the microbes for metabolic purpose (McKinley, 2001).

Depending on the environmental conditions, several redox processes can be used by microorganisms. In the presence of oxygen, aerobic oxidation of a reduced form of organic carbon (ROC), here generally written as CH_2O, is oxidized by oxygen. In this aerobic process, the electrons from the oxidation of the ROC are used by oxygen being the electron acceptor by forming H_2O. Deeper down in the subsurface, free oxygen is absent and anaerobic processes are used by microbes such as denitrification, Mn(IV) or Fe(III) reduction, sulfate reduction and methane production (Table 2.1). In these processes the electron acceptors are NO_3^- forming N_2 or NH_4^+, Mn(IV) forming Mn(II), Fe(III) forming Fe(II) or SO_4^{2+} forming HS^- or S^{2-}. These electron acceptors are used in a predictable sequential series, according to the free energy yielded by their reduction (DeLong, 2004). The ROC can be, for example, fatty acids or low molecular weight organic acids (such as acetate) deriving from deposited organic matter and transported by the slow groundwater flow through the sediments or hydrocarbons in oil reservoirs. Several investigations on nutrient availability and microbial activity within

2. Background

Reaction	G (kJ/mol)	type
$CH_2O + O_2 \rightarrow CO_2 + H_2O$	-475	aerobic respiration
$5CH_2O + 4NO_3^- \rightarrow 2N_2 + 4HCO_3^- + CO_2 + 3H_2O$	-448	denitrification
$CH_2O + 3CO_2 + H_2O + 2MnO_2 \rightarrow 2Mn^{2+} + 4HCO_3^-$	-349	Mn(IV) reduction
$CH_2O + 7CO_2 + 4Fe(OH)_3 \rightarrow 4Fe^{2+} + 8HCO_3^- + 3H_2O$	-114	Fe(III) reduction
$2CH_2O + SO_4^{2-} \rightarrow H_2S + 2HCO_3^-$	-77	sulfate reduction
$2CH_2O \rightarrow CH_4 + CO_2$	-58	methane production

Table 2.1.: *Redox reactions providing energy (G) for microorganisms, modified after McKinley (2001).*

deep sediments support the theory that recalcitrant organic matter becomes reactivated during burial by rising temperature. This can lead to an increasing nutrient availability in deeper sedimentary succession stimulating microbial activity in the depth (Wellsbury et al., 1997; Horsfield et al., 2006; Parkes et al., 2007).

The previously mentioned processes are not truly independent from photosynthesis as long as the source of the reduced organic carbon is fossil organic matter, that was once synthesised using the energy of the sunlight. In sediments that contain only little or no buried organic matter (e.g. granites and basalts), so-called subsurface lithoautotrophic microbial ecosystems (SLiMEs) are able to synthesise small organic compounds such as acetate (acetogenesis) or methane (methanogenesis) from inorganic sources. This ecosystem uses the so-called "geo-gas" (deriving from deep crust of the Earth) containing hydrogen as reducing component and carbon dioxide as oxidising complement (Pedersen, 2000). The products can be further used by other microorganisms and, therefore, these microbial communities are truly independent from sunlight (Figure 2.3).

The CO_2 within the deep Earth is formed by subsolidus decarbonation reactions of carbonates or oxidation of organic carbon in the depth (Santosh and Omori, 2008a,b). In addition, Pedersen (2000) quotes six possible

2.1. Microorganisms and the Deep Biosphere

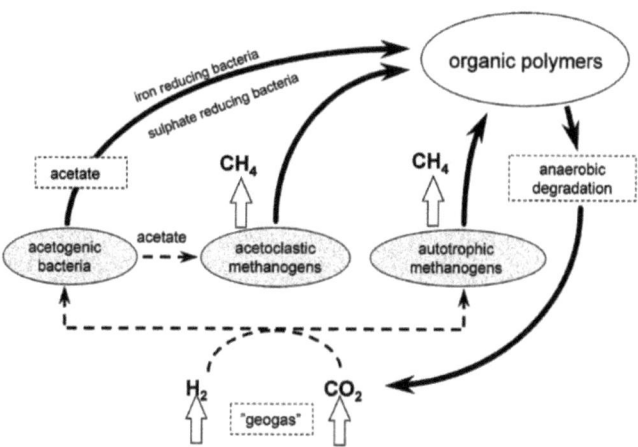

Figure 2.3.: *The deep hydrogen-driven biosphere hypothesis, illustrated by its carbon cycle. At relevant temperature and water availability conditions, intraterrestrial microorganisms are capable of performing a life cycle that is independent of sun-driven ecosystems. Hydrogen and carbon dioxide from the deep crust of the Earth are used as energy and carbon sources. Figure taken from Pedersen (2000).*

processes whereby hydrogen may be generated in the Earth's crust: (1) the reaction between dissolved gases in the carbon-hydrogen-oxygen-sulfur system in magmas, especially in those with basaltic affinities, (2) the decomposition of methane to carbon (graphite) and hydrogen at temperatures above 600°C, (3) the reaction between CO_2, H_2O and CH_4 at elevated temperatures in vapors, (4) radiolysis of water by radioactive isotopes like uranium and thorium and their daughter isotopes, and potassium (^{40}K), (5) cataclasis of silicates under stress in the presence of water, and (6) hydrolysis by ferrous minerals in mafic and ultramafic rocks.

Utilising the energy from redox reactions and mediating the decomposition and the production of organic molecules diversely composed microbial communities are the main engines driving the Earth's biogeochemical cycles. Often microbes use nearly identical pathways forward or reverse

2. Background

to maintain these cycles. Within the carbon cycle, methane is formed by methanogenic *Archaea* from the reduction of CO_2 or acetate with H_2, but when the hydrogen tension is low (occurring when there is an association with hydrogen consuming sulfate reducers), the reverse process becomes favorable and methane is oxidised anaerobically by similar *Archaea* (Falkowski et al., 2008). In a similar way, acetate is oxidised stepwise into CO_2 in the citric cycle, used by green sulfur bacteria and some Archaebacteria to assimilate CO_2 into organic matter (Wächtershäuser, 1990).

Of further importance for microbial life is the cycle of nitrogen due to the synthesis of proteins and nucleic acids for microbial DNA/RNA. The biological process that helps to make N_2 accessible for the formation of proteins and nucleic acids is the reductive transformation of N_2 to NH_4^+. It is a biological irreversible process catalysed by the enzyme nitrogenase. The presence of oxygen inhibits this process and NH_4^+ is oxidised to NO_2^- and further, to NO_3^- by a suite of nitrifying bacteria (Falkowski, 1997). The nitrifiers also use small differences in the redox potential of the oxidation reaction to reduce CO_2 to reduced organic matter. In the absence of oxygen, further microbes use NO_2^- and NO_3^- as electron acceptors in the anaerobic oxidation of organic matter and thereby form N_2 in a respiration process closing the N-cycle. Overall it seems that the niches for nearly all possible redox reactions may be occupied by microbial metabolism.

In 2008, Chivian et al. reported the finding of a single-species ecosystem within deep Earth by sequencing DNA of a complete genome collected from a low-biodiversity fracture water from 2.8 km depth in a South African gold mine (Chivian et al., 2008). The genome of this bacterium, *Candidatus Desulforudis audaxviator*, indicates a motile, sulfate-reducing, chemoautotrophic thermophile that can fix its own nitrogen and carbon. This bacterium is capable of an independent life-style prepared to long-term isolation from the photosphere deep within the Earth's crust.

In marine and terrestrial sediments, buried organic matter provides a suitable source for carbon and energy for microbial metabolic processes.

2.2. Sedimentary macromolecular organic matter - composition, classification and maturation

This buried organic matter is found in sediments either dispersed, for example as kerogen, or accumulated in oil reservoirs or coal beds.

Up to now, the individual reactions that enable life on Earth are only insufficiently understood and discovering and describing the processes involved is a challenging task in modern microbiological and biogeochemical research. A deeper understanding of these processes is important for the survival of humans continuing to influence the fluxes of matter and energy on a global scale. Microbial life can easily live without humans, but humans however, cannot survive without the global environmental transformations provided by microorganisms (Falkowski et al., 2008).

2.2. Sedimentary macromolecular organic matter - composition, classification and maturation

When organisms die, the organic matter undergoes divers biotic and abiotic alteration processes, depending on its composition and on the environmental conditions during sedimentation. While the main part of the organic material (OM) is rapidly degraded and used as food source by organisms, part of the OM survives this biological cycle and enters the geological cycle by forming or being already complex macromolecules that undergo diverse alteration processes during burial. Large coal beds arose from terrestrial plant material preserved by water and mud from oxidation processes. They form huge compact layers with high accumulated organic matter content. Marine and lacustrine sedimentation of OM can lead to the formation of oil source rocks that contain dispersed macromolecular organic remains, kerogen.

2.2.1. Kerogen - dispersed macromolecular organic matter

The term kerogen (*greek*: keros = wax) was first used by Crum Brown in 1912 for the description of the organic matter of a Scottish oil shale that produced a waxy oil upon distillation (Vandenbroucke and Largeau, 2007). Thus, the first definition of kerogen as a material generating oil and gas upon high temperature heating was restricted to OM rich rocks of economic importance. Later the definition was extended to all OM in rocks that is able to generate oil and gas (Trager, 1924). Using the term 'organic matter' includes both kerogen that has already generated oil and the generated oil itself. Therefore, a distinct definition of kerogen was needed being more orientated to the structural and compositonal characteristics of kerogen.

In the modern definition, kerogen is the macromolecular organic matter which is insoluble in common organic solvents (Durand, 1980) in contrast to the soluble part of the OM which is termed bitumen. This insoluble macromolecular organic matter is originated from plant material or algae modified by diverse microbial and thermal alteration processes during burial. Therefore, kerogen represents an important part within the global carbon cycle. It is a long time sink for Earths carbon due to the fact that the geological cycle lasts millions of years. Kerogen represents the largest OM pool on Earth including 10^{16} tons of carbon compared to approximately 10^{12} tons in living biomass (Durand, 1980). It consists of condensed aromatic units and aliphatic chains linked by hetero atoms mainly oxygen and sulfur. Depending on its origin and structural composition, kerogens can be divided into at least three different types: Types I, II and III kerogen.

Sources and environmental composition of dispersed organic matter

The starting materials from which kerogen forms are the primary producers that contributed to the sedimentary organic matter. In Earth's history, the first contributors were algae followed later by terrestrial higher plants emerging during the late Silurian and Devonian period. Primary producers

2.2. Sedimentary macromolecular organic matter - composition, classification and maturation

are photoautotrophic organisms that transform CO_2 from the atmosphere into their own organic matter by using the energy of the sunlight for photosynthesis. Therefore, the habitat for primary producers is restricted to the land surface and the upper layer of the water column. In aquatic environments, phytoplankton is one of the largest contributors to the sedimentary OM. Main contributors to phytoplankton are diatoms, dinoflagellates, coccolithophorids, green algea and cyanobacteria.

Sedimentation is the process when particles settle down through the water column to the seafloor and the bottom of rivers or lakes and, therefore, occurs only in aquatic environments. Terrestrial plant material accumulates during transportation to aquatic systems by wind, erosion and rivers. The deposited organic matter undergoes various biotic and abiotic degradation processes under oxic conditions. Only very inert substances like waxes or cutan can partly escape this degradation. It is only about 0.1% to 1% of the living biomass of the source organisms that is incorporated into the sedimentary matter worldwide (Tissot and Welte, 1984b).

During sedimentation in lacustrine or marine environments diverse microorganisms rework the OM. The microbial material is also incorporated in the organic debris provided by the primary producers. Sediments contain organic matter in various amounts depending on the amounts of primary producers, degradation and reworking during sedimentation. Thus, the amount of OM deposited on the seafloor correlates with the environmental conditions. The balance between organic productivity (limited by nutrient availability) and oxidative degradation (controlled by thickness of the oxic zone; Demaison and Moore, 1980), variations in productivity by climatic changes and varying preservation due to varying atmospheric CO_2 content (Berner, 1994) is outlined in the literature.

2. Background

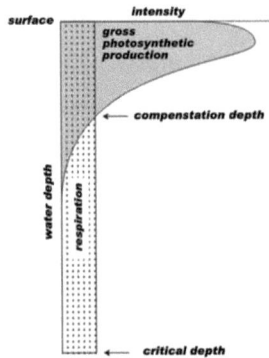

Figure 2.4.: *The photosythetic production by phytoplancton decreases with decreasing light intensity but respiration appears throughout the whole water column. At the compensation depth phytoplanctonic production and respiration intensity is equal. The critical depth is reached when all phytoplanktonic matter is respirated. Taken from Killops and Killops (2004)*

Figure 2.4 shows the relationship between productivity from primary producers and respiration of the OM within the water column. During the upper few meters, where sunlight is present, high photosynthetic production by phytoplankton is observed, decreasing with increasing depth and decreasing light intensity. At the compensation depth, the rates of fresh production and respiration are equal. Below this point respiration of the OM dominates. The critical depth is reached when the total respiration in the water column above equals the gross primary production. Generally it can be said that the amount of OM that reaches the seafloor depends on the depth of the water column. The deeper the water is, the longer is the sinking time and the more degradation and respiration will take place. The amount of OM that reaches the seafloor decreases with approximately a factor of 10 when water depth increases tenfold (Killops and Killops, 2004). According to *Stokes' Law* (Stokes, 1901), the terminal velocity (or sinking velocity) is given by the drag force (or frictional force) and the force by gravity. The latter force is influenced by the mass density of the spherical object and the fluid respectively, the gravimetrical acceleration and the radius of the spherical object while the drag force depends on the fluid's viscosity, the particles velocity and the radius of the spherical object. Therefore, the size of the particles, their density and their radius as well as the density of the water influences the sinking time and thus, the extend of degradation. Furthermore, possible adsorption to mineral matrix can influence these parameters. When the remaining particles are settled down

2.2. Sedimentary macromolecular organic matter - composition, classification and maturation

to the bottom, further microbial degradation takes place, depending on the rate of sedimentation and the amount of oxygen present in the surrounding sediment. Along with ongoing burial, depending on sedimentation rate, alteration of the OM continues, leading to dispersed OM in sediments. Within the sediments the degradation of the OM continues by anaerobic microorgansims that gain their energy from oxidising the sedimentary OM as described in Chapter 2.1.4 and in greater depth at higher temperatures, by thermal cracking processes described below.

Classification of kerogen

The classification of organic matter into different kerogen types is based on the composition of organic matter being the result of source material and different depositional and environmental conditions. The first studies on kerogen description were reported by Down and Himus (1941) and later by Forsman and Hunt (1958). They distinguished the organic matter into two main types, one was organic matter from coal deposits and the other type showed a higher aliphatic character. Furthermore, they revealed that elemental changes occur during maturation upon burial. The modern basic classification of kerogen recognises three main types. Based on the elemental composition of kerogen (Durand and Espitalié, 1973; Tissot et al., 1974) and comparison of this data set with related literature data, three main types were defined by plotting the atomic ratios H/C against O/C in a diagram used by van Krevelen (1961) for the description of coal maturation.

In Figure 2.5, the three types of kerogen are plotted in a van Krevelen diagram (a). Maturation of the organic matter is indicated by decreasing O/C and H/C ratio during the evolution upon burial. The low mature Type I kerogen starts with a high H/C ratio due to high amounts of aliphatic constituents, the low mature Type III kerogen starts with low H/C and high O/C ratios due to its high amounts of phenolic constituents. The Type II

2. Background

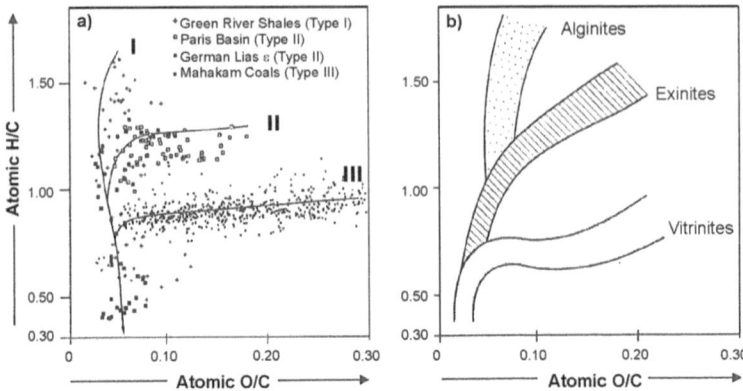

Figure 2.5.: *Evolution path of the three different kerogen types plotted in a van Krevelen type diagram (a) from Type I starting at low maturity with high H/C and low O/C atomic ratios to Type III starting at low maturity with low H/C and high O/C atomic ratios. Closely related is the evolution of coal macerals (b) upon geological burial plotted in a diagram of H/C vs O/C ratios (van Krevelen, 1961), adapted from Tissot and Welte (1984b); Vandenbroucke and Largeau (2007)*

kerogen is located in the middle between Type I and Type III. The same diagram type was initialy used to outline the evolution pathway of coal macerals (b) (van Krevelen, 1961). The coal macerals describing distinct organic entities within coals are described in detail below (Chapter 2.2.2). Due to its aliphatic constitution, alginite starts with high H/C values while the maceral vitrinite deriving from higher terrestrial plants starts with a high O/C ratio. Exinite, later renamed liptinite, is located in the middle between alginite and vitrinite.

Kerogen Type I

The reference Type I kerogen is found in the Green River Shale (GRS), deposited during Eocene in a large area at the watershed of the present-day

2.2. Sedimentary macromolecular organic matter - composition, classification and maturation

Green River, a tributary of the Colorado River around the Unita Mountains of northeastern Utah. Three independent intermontane basins resulted from uplift of the Rocky Mountains and were filled with sediments from the surrounding deposited in the intermontain lakes containing sandstones, mudstones, siltstones, limestones, oil shales (the Green River Shale) and coal beds, as well as volcanic ash layers from the Absaroka volcanic field in the north. The GRS is the source rock of the primary petroleum system in the Uinta Basin (UT, USA). The immature GRS kerogen is highly aliphatic with H/C ratios >1.5 (Figure 2.5 a). It has a low oxygen content, immature samples of the GRS have O/C ratios lower than 0.1 (Tissot and Welte, 1984b). Compared with the other kerogen types, it contains only a low content of polyaromatic rings. The small amount of oxygen is mainly found in ether groups (Fester and Robinson, 1966), although esters are also present (Figure 2.6). In pyrolysis experiments up to 600°C, this kerogen produces a larger yield of volatile compounds and, therefore, a higher yield of oil during maturation than other kerogen types. Organic solvent extracts as well as the generated oils and the pyrolysis products of this kerogen are rich in long chain n-alkanes with chain length up to more than 40 carbon atoms. In kerogen Type I, the paraffinic hydrocarbons dominate over cyclic structures (Figure 2.7A). The high amounts of lipids result either from accumulation of algal remains mainly *Botryococcus* (Marchand et al., 1969), or from biodegradation during deposition in lacustrine environments. Although Type I kerogen has the highest potential for oil generation among all OM types, it is estimated that it is the source of only 2.7% of the world's reserves of oil and gas (Klemme and Ulmishek, 1991).

Other examples of Type I kerogen are the lacustrine kerogen of Autum boghed coal (France) and Torbanite (Scotland) as well as Coorgonite (Australia) and the marine Type I kerogen of the organic matter rich Tasmanite from Tasmania (Australia) (Hutton, 1987; Revill et al., 1994).

2. Background

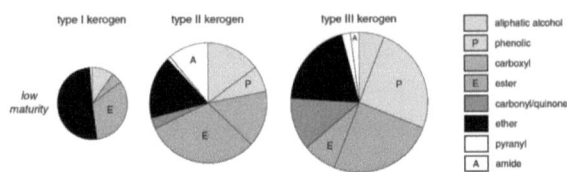

Figure 2.6.: *Distribution of oxygen containing functional groups in the three different types of kerogen in low maturation stage, adapted from Killops and Killops (2004). The oxygen in Type I kerogen is mainly located in ether and ester groups, in Type II kerogen mainly in ester, ether and carboxyl groups and in the terrestrial Type III kerogen mainly in phenols and carboxyl groups as well as in ether and ester groups.*

Kerogen Type II

Kerogen Type II is found in many OM-rich sediments, including the lower Toarcian shales of the petroleum system from the Paris Basin and the equivalent Posidonia Shale (Liassic ϵ) in Germany. Kerogen Type II is the source material of a vast number of oil and gas fields, e.g. the Devonian and Colorado Group of Cretaceous Age from Western Canada, Paleozoic sources from North Africa, Jurassic source rocks of Western Europe and Saudi Arabia and some Cretaceous and Tertiary source rocks of West Africa. In the immature kerogen, the H/C atomic ratio is relatively high and the O/C atomic ratio is low, 1.3 and 0.15 respectively (Figure 2.5). Polyaromatic units are more abundant than in Type I, but less than in Type III. Oxygen is mainly bound in ester bonds and carboxylic acid groups (Figure 2.6). Beside a large amount of aromatic components, this kerogen type contains many cyclic aliphatic moieties that became aromatised within progressive maturation (Vandenbroucke and Largeau (2007) and references therein). The extracts and pyrolysates of this kerogen yield lower amounts of products than Type I. Kerogen Type II is usually related to marine sediments. The primary source for this OM is a mixture of phytoplankton, zooplankton and bacterial matter of intense rework. Beside oxygen, sulfur is also present, being part of heterocycles and of sulfide bonds (Tissot and Welte, 1984b).

2.2. Sedimentary macromolecular organic matter - composition, classification and maturation

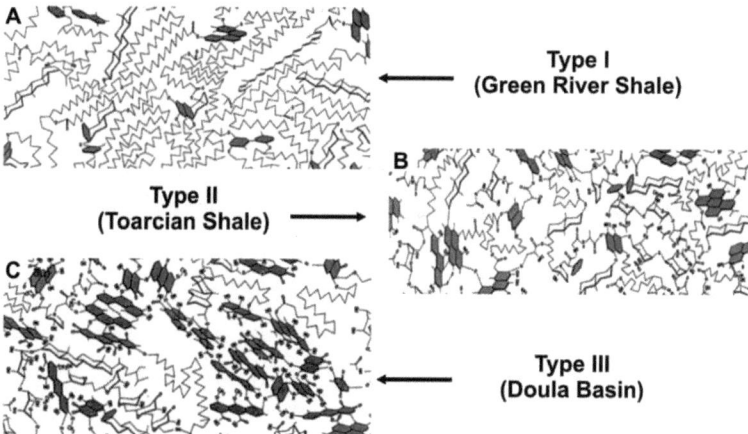

Figure 2.7.: *Chemical structures of the three kerogen types at the beginning of diagenesis, after Béhar and Vandenbroucke (1987). A: immature kerogen Type I (Green River Shale) with predominantly long chain aliphatic structures, B: immature kerogen Type II (Toarcian Shales) with aromatic ring structures and aliphatic structures, C: immature kerogen Type III (Doula Basin) with high content of cyclic and aromatic structures.*

The term Type II-S can be used for sulfur rich Type II kerogens. These kerogens can be grouped to a Type II subgroup containing 8-14% (by weight) organic sulfur resulting in a S/C atomic ratio higher than 0.04 (Orr, 1986).

Kerogen Type III

In Tissot et al. (1974), the reference Type III kerogen is based on the Upper Cretaceous sedimentary rocks of the Douala Basin (Logbaba, Cameroon) though, of course, humic coals also fall in this category (Durand et al., 1977) as discussed further below. These sediments were deposited in a shallow marine environment, sourced by paleo deltaic inputs of the Niger River (Albrecht et al., 1976). The immature kerogen has a low H/C ratio of usually below 1.0 (e.g. 0.8 for kerogen from the Doula Basin (Durand and Espitalie, 1976) and a high O/C ratio up to 0.2 or 0.3 (Tissot and Welte,

1984b). Kerogen Type III is often found in deltaic settings and in thick detrital sedimentation along continental margins. Contributor to this type of OM is mainly higher plant material that is often highly reworked during transportation. Only resistant chemical components of this terrestrial organic matter escape degradation and reache the site of sedimentation. Microbial degradation within the sedimentary basin is often limited due to fast sedimentation, rapid burial and high subsidence rates on continental margins.

Due to its origin from higher plant material, this kerogen type comprises high amounts of polyaromatic units as well as ketones and carboxylic acids. A minor amount of oxygen is found in ester functions (Figure 2.6). Some non carbonyl oxygen is located in ether bonds. Aliphatic structures are of minor importance but appear in varying amounts. In general, these structures consist of a few long chains of more than 25 carbon atoms originating from higher plant waxes and some chains of medium length of 15 to 20 carbon atoms steming from vegetable fats as well as some methyl groups and other short chain aliphatics. The kerogen Type III does not form any oil generating shales, but, if buried deeply enough, can provide a gas source rock.

Additional examples for the occurrence of kerogen Type III can be found in the lower Manville shales in Alberta (Canada) and in some Tertiary deltas such as the Mahakam delta (Indonesia).

It is important to note that kerogens may plot in the van Krevelen diagram (Figure 2.5) between the classical evolution pathway outlined in the van Krevelen diagram for different reasons. These variabilities derive from different constitution of the contributing organisms and mixing of marine and terrestrial material can also cause variability. Therefore, some marine kerogens show unusual high O/C ratios due to additional supply of higher plant debris. This was found e.g. for the Cretaceous Viking shales in Alberta or some Cretaceous black shales in the North Atlantic.

2.2. Sedimentary macromolecular organic matter - composition, classification and maturation

Maturation of kerogen organic matter

During burial, ongoing alteration of the organic remains takes place, visible by changes in the elemental composition shown in the van Krevelen diagram (Figures 2.5a and 2.8). Mainly during the first evolutional stages of the OM transformation (diagenesis), complex macromolecules are formed due to a complex process involving condensation and addition reactions of smaller compound units, and also by various degradation reactions. The degradation of the OM can be caused by biotic and abiotic alteration (in the upper part of the sediment) or abiotic cracking of bonds by thermal processes (in deeper sediments) (Tissot and Welte, 1984b; Vandenbroucke and Largeau, 2007).

According to changes in the O/C and H/C atomic ratios of the kerogen during ongoing maturation, first loss of oxygen, then hydrogen, the evolution of the OM can be separated into different maturity stages. The generation of free hydrocarbons (oil and gas) leads to changes in the constitution of the kerogen by the loss of aliphatic structures. The generation of hydrocarbons as well as ring condensation reactions and dehydrogenation lead to an increase of the content of aromatic structures within the kerogen. In addition to elemental composition, the level of maturity of the kerogen can be appraised by reflected light microscopy using optical properties that are correlated to aromatisation of the organic matter with increasing maturation. This parameter is mainly influenced by the maceral vitrinite and is, therefore, called vitrinite reflectance (R_0) (*cf.* Chapter 2.3.2). Another useful maturation parameter for kerogen is the T_{max} value, the temperature where a kerogen produces the highest amounts of hydrocarbons in Rock-Eval pyrolysis experiments (*cf.* Chapter 2.3.3). With ongoing maturation, kerogen shows a higher stability upon pyrolysis. The less heteroatoms and aliphatic chains a kerogen contains, the more stable this kerogen is upon pyrolysis. Thus, a higher T_{max} value indicates a more stable and inert kerogen. During maturation of the kerogen, the T_{max} value increases, reflecting the formation of more and more stable macromolecules by releasing labile bound structure

2. Background

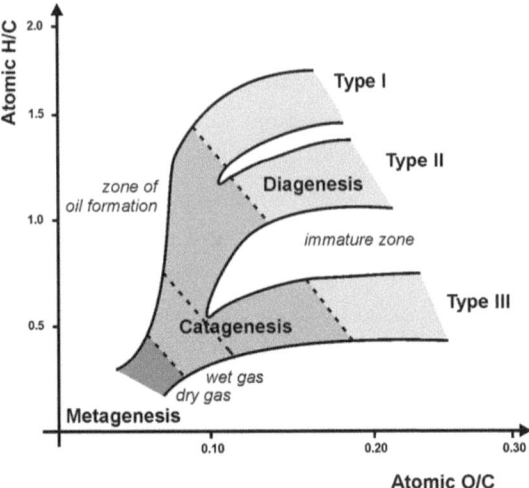

Figure 2.8.: *Scheme of kerogen evolution showing diagenesis, catagenesis and metagenesis stage, including oil and gas generation zones, modified after Tissot and Welte (1984b).*

units, for example hydrocarbons during petroleum generation.

The first definition of maturation zones for organic matter was made by Vassoevitch et al. (1969). The zones named (by Vassoevitch) protocatagenesis, mesocatagenesis and apocatagenesis were later renamed by Tissot and Welte (1984b) using vitrinite reflectance microscopy (*cf.* Chapter 2.3.2) to diagenesis ($R_0 < 0.5\%$), catagenesis (R_0 0.5-2.0%) and metagenesis ($R_0 < 2.0\%$) (Figures 2.8 and 2.12).

Determination of the loss of oxygen and hydrogen is a well established method for describing coal maturation because these are the predominant elements in terrestrial OM beside carbon. When describing marine kerogen, the element nitrogen is of greater importance. This led to the definition of a specific first stage of the OM evolution, the early diagenesis (Vandenbroucke and Largeau, 2007).

2.2. Sedimentary macromolecular organic matter - composition, classification and maturation

Early diagenesis

Early diagenesis occurs only at a very restricted depth. In this zone of Om transformation the most reactive compounds in the nitrogen rich marine sediments are degraded. These changes can be observed by a decrease of the N/C atomic ratio. The N/C ratio in terrestrial plants is generally low (<0.05 for a given H/C ratio of approximately 1.0). However, planktonic organisms show higher N/C ratios (>0.08 fo a H/C ratio of 1.3 to 1.5). These ratios generally depend on the amino acid content of the contributing source organisms (Vandenbroucke and Largeau, 2007). Usually, early diagenesis appears only in the water column and in the upper surface sediments down to a few meters below surface. The main reaction pathway for degradation of nitrogen containing compounds is the elimination of amino acids and ammonia by hydrolysis (Schnitzer, 1985). In very immature sediments being only a few thousand years old (e.g. the algal sapropel from the Mangrove Lake (Bermuda)) nitrogen is mainly found in amide groups (Knicker et al., 1996). Older, but still immature, algal sediments contain the remaining nitrogen predominantly in the form of pyrroles. The end of early diagenesis is reached when nitrogen in the OM is no longer hydrolysable. Only in late stages of thermal methane generation (dry gas) nitrogen will be released again as N_2 gas (Littke et al., 1995; Gillaizeau et al., 1997).

In marine sediments, the incorporation of sulfur in the sediments is observed during early diagenesis, forming sulfur rich kerogens (Type II-S). Some highly sulfur rich kerogens of 15 wt% sulfur (S/C ratio 0.10) have been reported in marine sediments of the Miocene Monterey Formation (Sinninghe-Damsté et al., 1989). Sulfur incorporation can happen abiotically by incorporation of pyrite microcrystals and pyrite framboids, as well as biotically in anaerobic marine sediments by sulfate reducing bacteria (SRB) belonging mostly to the *Desulfovibrio* genus (Berner, 1984). The initial step of the pyrite formation is the bacterial reduction of sulfate to H_2S using the sedimentary organic matter as reducing agent (*cf.* 2.1.4). Pyrite then forms during shallow burial by the reaction of the H_2S with detrital iron minerals

2. Background

leading to a series of metastable iron monosulfides which transform to pyrite during early diagenesis (Berner, 1984).

Diagenesis

The stage of diagenesis is the maturation phase of the main decrease of oxygen as indicated in the van Krevelen diagram by the main decrease of the O/C atomic ratio (Figure 2.8). The oxygen decrease correlates with an increase of carbon content in the sediment. However, the decrease of H/C ratio is rather small. The end of diagenesis does not correlate to the end of O/C atomic ratio decrease which is still observed during early stages of catagenesis. Diagenesis can occur over a range of more than one kilometer depth but describes the alteration processes during the early stages of burial at relatively low temperature and pressure. The main degradation and alteration processes during diagenesis are biotic. During diagenesis, the organic carbon is oxidised mainly to CO_2 by microorganisms using sulfate, nitrate, Mn(IV) oder Fe(III) as oxidising agent and gaining their energy from the Gibbs free energy of these redox reactions (*cf.* Chapter 2.1.4). In addition, some abiotic processes are possible by catalytic reactions e.g. on mineral surfaces (Bada and Mann, 1980; Tegelaar et al., 1989; Killops and Killops, 2004). Along with increasing burial, the consolidation of the organic matter increases. The moisture content decreases while temperature and pressure increase with depth. The end of diagenesis and beginning of catagenesis correlate with the beginning of oil formation corresponding to temperatures higher than 60°C but usually close to 100°C (Killops and Killops, 2004). At the end of diagenesis, the rising temperature limits the microbial activity. At this stage of maturation, the thermal energy becomes sufficiently high to induce the break of chemical bonds within the kerogen. Methanogenesis is the last well described biological process that is found to operate in the depth at high temperatures of up to 75°C (Killops and Killops, 2004). Since it was recognized that thermophilic bacteria can be extremely adaptable to extreme environments surviving at relatively high temperatures

2.2. Sedimentary macromolecular organic matter - composition, classification and maturation

(cf. Chapter 2.1), it is conceivable that bacterial activity is not restricted to diagenesis but is still observable at the early stages of catagenesis.

Catagenesis

Catagenesis is the name for the maturation phase of the main decrease of hydrogen content in the organic matter and, therefore, the stage of petroleum generation. For a Type II kerogen, the H/C ratio decreases from around 1.2 to 0.5, indicating the release of hydrocarbons. It is the beginning of the so-called oil window. Temperature is high enough to induce the cracking of chemical bonds, and the generated hydrocarbons of different carbon chain length start to migrate out of the source rock into carrier systems until they are caught under a so-called cap rock (a non permeable layer building a seal) to form an oil reservoir (secondary migration) or they reach the surface (tertiary migration, e.g. oil leakage on the seafloor offshore California; Mangelsdorf and Rullkötter, 2003). Kerogen Type I and II are rich in aliphatic structures and have, therefore, the highest oil potential on a carbon normalised basis. If the aliphatic rich biomacromolecules cutan or suberan were once incorporated in the sedimentary kerogen, there is a high potential of oil generation. This is also a reason for the potential of the some coals, for example those found in south-east Asia, Australia and New Zealand, to generate highly paraffinic and waxy oils (Killops et al., 1998; Powell, 1991).

The end of catagenesis is given at a H/C ratio of about 0.5 and a vitrinite reflectance of about 2% (Tissot and Welte, 1984b). At this maturation stage, all types of kerogens plot in the same region of the van Krevelen diagram and no more distinction between the different kerogen types can be made. A structural model of kerogen in different stages of maturation was proposed by Béhar and Vandenbroucke (1987). Figure 2.9 shows the theoretical structures for a kerogen Type III according to different H/C and O/C ratios. At the beginning of diagenesis (A), the kerogen consists of aromatic and

2. Background

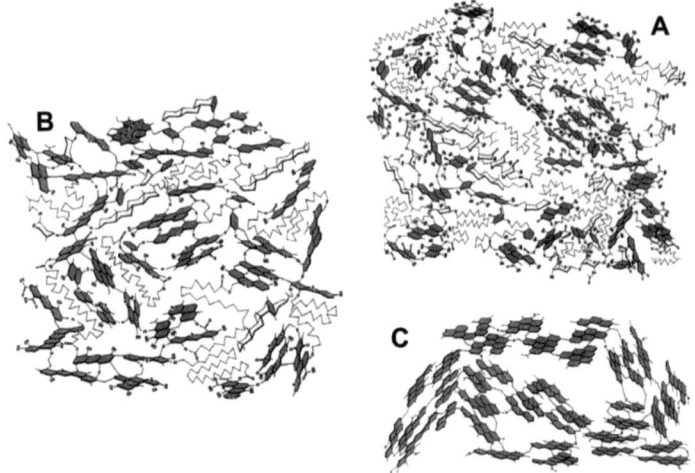

Figure 2.9.: *Theoretical structures of a Type III kerogen at different maturity stages. A: at the begin of diagenesis, $H/C = 1.06$, $O/C = 0.281$; B: at the end of diagenesis, $H/C = 0.98$, $O/C = 0.138$; C: at the end of catagenesis, $H/C = 0.67$, $O/C = 0.059$, from Béhar and Vandenbroucke (1987).*

aliphatic structural units with high amounts of oxygen, indicated by a high O/C atomic ratio. The decrease of this atomic ratio at the end of diagenesis (B) and at the end of catagenesis (C) is accompanied by a relative increase of aromatic structures within the kerogen. At thermal maturity higher than 1.5% vitrinite reflectance, the generation of gas (methane) from secondary cracking of hydrocarbons can be observed (Dieckmann et al., 1998).

Metagenesis

During the stage of metagenesis, reorganisation of the residual aromatic network of the kerogen occurs. The aromaticity continues to increase and the only hydrocarbon still generated from the kerogen is methane (dry gas) by thermal cracking and elimination of methyl groups (Béhar, 1997). Part of the generated methane may also derive from secondary cracking of already

2.2. Sedimentary macromolecular organic matter - composition, classification and maturation

generated higher hydrocarbons. In addition to methane, some hetero elements are also released in the form of CO_2, H_2S and N_2 (Béhar et al., 2000). Although there are no compositional differences between the three kerogen types at this level of maturity, the size of the aromatic clusters in the macromolecule depends on the former kerogen type, increasing from Type III to Type I (for Type III: around 50 Å, Type II: 100-200 Å, Type I: up to 1000 Å; Oberlin et al., 1980). In metagenetic maturity stage, the aromatic character of the macromolecules increases due to ring condensation reactions. This evolution can be observed using electron paramagnetic resonance (EPR) spectroscopy. Plotting the paramagnetic susceptibility against the H/C ratio, acting as a maturity parameter, a decrease of free radicals in the kerogen is observable during ongoing metagenesis due to radical recondensation. In contrast, during previous catagenesis the number of free radicals usually increases due to the high rates of thermal cracking processes (Marchand and Conard, 1980).

2.2.2. Coal - accumulated terrestrial macromolecular organic matter

In addition to high amounts of dispersed OM in sediments like shales, clays or sandstones, a notable amount of sedimentary organic matter is found in a concentrated form as coals. Coals are sedimentary rocks formed from mainly terrestrial plant material and especially from lignocellulosic starting material (wood, roots, bark) deposited in swamps, preserved by water and mud from oxidative processes. Different types of coals are distinguished ranging from peat to anthracite according to their maturity rank. Beside some minor contribution of inorganic mineral matter, coals consist, therefore, mainly of kerogen and bitumen. Schopf (1956) defines coal as "readily combustible rock containing more than 50% by weight and more than 70% by volume of carbonaceous material, formed by the compaction or induration of variously altered plant remains" and the International Commission

2. Background

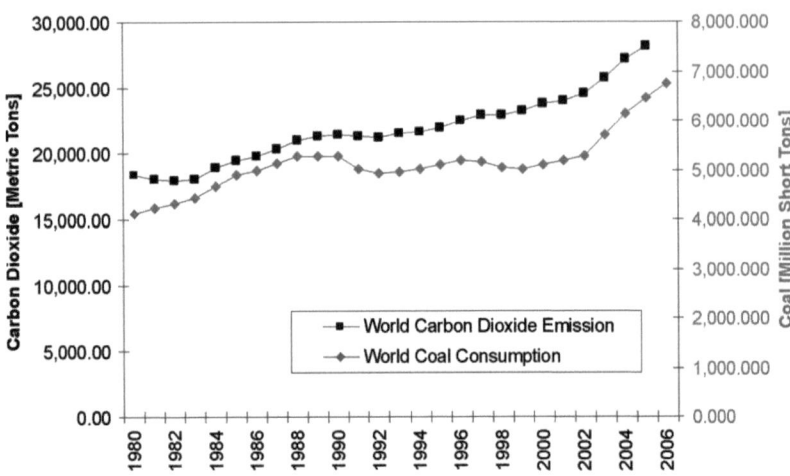

Figure 2.10.: *World coal consumption (red) in million short tons (1 short ton = 2000 lb = 907.18 kg) and world CO_2 emission (black) in metric tons. Data taken from Department of Energy (DOE) / Energy Information Administration (EIA) - International Energy Annual 2006. Source: http://www.eia.doe.gov/international/*

for Coal Petrology (ICCP) defines coal as "a combustible sedimentary rock formed from plant remains in various stages of preservation by processes which involved the compaction of the material buried in basins, initially of moderate depth" (ICCP, 1963).

Coal is a fossil fuel used mainly for energy production by combustion in coal power stations worldwide. Because it is so plentyful, coal is the most important energy reserve. The Department of Energy (DOE) of the US Government estimates the world coal reserves at 930 billion tons[1]. Coal deposits occur everywhere in the world located in sedimentary basins. The coal beds (seams) are typically sandwiched between sandstones or shales.

---

[1]Source: Department of Energy (DOE) / Energy Information Administration (EIA) of the U.S. Government, Report: DOE/EIA-0484(2008), Release Date: June 2008 - http://www.eia.doe.gov/oiaf/ieo/coal.html.

2.2. Sedimentary macromolecular organic matter - composition, classification and maturation

Commercially mined coal beds are found in Europe, Asia, Australia and North America (Landis and Weaver, 1993). Burning of fossil fuels in order to satisfy the demand of energy is one of the crucial factors of the emission of the greenhouse gas CO_2 (Figure 2.10), being the product of the oxidation process. Intense global mining and burning of coals causes an unnatural intervention in the geological carbon cycle and influences its global equilibria. Because global warming has been linked to emitted greenhouse gases, alternative energy production techniques as well as CO_2 sequestration and recycling techniques, are being implemented.

Sources and environmental composition of coal

When woody material is deposited in swamps, the organic material is preserved in an anoxic environment. In this special case, the oxidative degradation and consumption of OM is prevented. Anaerobic microbial degradation and compression of the sedimentary matter leads to thick layers containing accumulated organic matter, the total organic carbon (TOC) in this formations is high, it can made up to 90%. The terrestrial plant remains form complex macromolecules. Thus, coals consist mainly of a terrestrial kerogen type. Originating from different plant material, the organic matter forming coals are summarised to macerals and maceral groups. The term macerals was proposed by Marie C. Stopes in 1935 and describes distinct organic entities composing coal related to different originating materials. Macerals can be seen as descriptive equivalents of the inorganic units composing rock masses called minerals (Stopes, 1935). There are three basic groups of macerals: (1) vitrinite, deriving from coalified woody tissue, (2) liptinite, deriving from waxy coatings of plants and (3) inertinite, deriving from charred and altered plant cell walls. They can be identified by transmitted or reflected light microscopy (*cf.* Chapter 2.3.2). Table 2.2 gives an overview about the macerals and its origin found in bituminous coals.

The main distributor to the molecular structure of terrestrial organic

2. Background

Maceral group	Maceral	Origin
vitrinite	telinite	cellular structure of wood, leaf and root tissue
	collinite	structureless, infilling gel
	vitrodetrinite	unidentified cell fragments
liptinite	sporinite	spore and pollen cases
	cutinite	waxy coating of leaves and stems
	suberinite	cork tissues, e.g. bark and root walls
	resinite	resin bodies
	alginite	algal tests
	liptodetrinite	unidentified liptinite fragments
	fluorinite	lenses/layers, possibly plant essential oils
	bituminite	wisps or groundmass, from lipids
	exudatinite	veins of expelled bitumen-like material
inertinite	fusinite	charred wood and leaf tissue
	semifusinite	partially charred wood and leaf tissue
	macrinite	charred gel material
	micrinite	charred liptinitic material
	sclerotinite	fungal remains
	inertodetrinite	unidentified intertinite fragments

Table 2.2.: *Petrographic classification of bituminous coal constituents (Bend, 1992) including maceral groups and related macerals and the originating material.*

matter is lignin. Lignin is mostly found in cell walls of higher terrestrial plants where it is forming a network around the cellulose fibers in the woody material of plants. It is a high molecular weight biopolymer from phenolic compounds formed by condensation reactions (Figure 2.11). The main building blocks of lignin are the phenolic alcohols coumaryl, coniferyl and sinapyl alcohol biosynthesised from glucose under enzymatic control (Nimz, 1974; Adler, 1977).

Other plant or organism remains are only minor contributors to the macromolecular organic matter. Waxes are mixtures of high melting con-

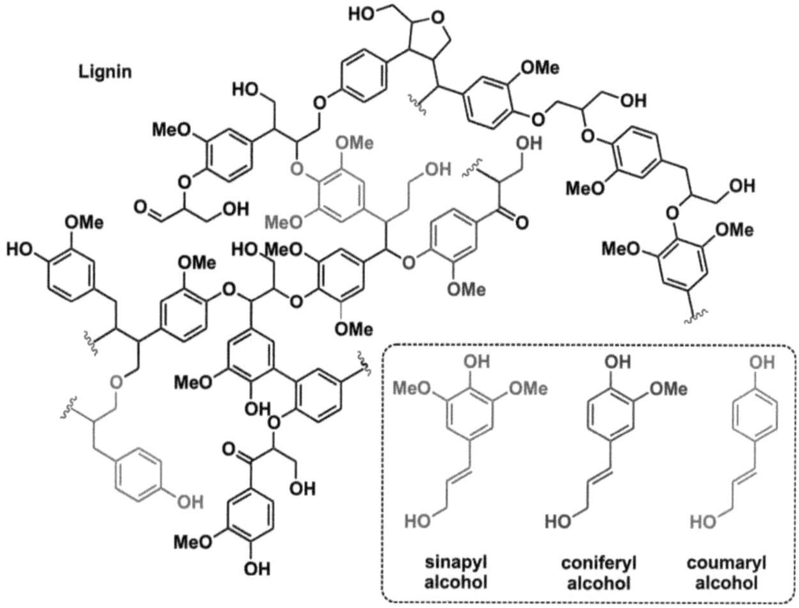

2.2. Sedimentary macromolecular organic matter - composition, classification and maturation

Figure 2.11.: *Random, partial structure of lignin after Nimz (1974) with colored main lignin building blocks coumaryl, coniferyl and sinapyl alcohol, Me=CH_3.*

stituents forming protective coatings e.g. on leaf cuticles. Important compounds are esters from fatty acids, condensed with long aliphatic saturated alcohols. These fatty acids and alcohols have chain lengths mainly in the range of C_{24} to C_{28}. Chain lengths with even numbers of carbon atoms dominate according to their biosynthesis from acetyl units. Similar compounds also used as protective coating are cutin (a polymer from hydroxy- and dihydroxy fatty acids) and suberin (a polymer from α, ω-dicarboxylic acids and ω-hydroxy fatty acids).

Another important compound class are the terpenoids. They appear in various structures in organisms and only some important are mentioned here. Monoterpenoids (with two isoprene units, e.g. pinene, menthol or camphor) are flavoring substances in plant oils due to their volatility.

2. Background

Triterpenoids (with 6 isoprene units) e.g. steroids (tetracyclic triterpenoid) or hopanoids (pentacyclic triterpenoids) are components of cell membranes most likely used as membrane rigidifiers. Among a various bunch of compounds, they can be used as biomarkers to indicate e.g. environmental conditions during sedimentation or the contribution of microbial biomass to the OM.

Lipids (defined as all the substances produced by organisms that are effectively insoluble in water but extractable by solvents that dissolve fats; Killops and Killops, 2004) can contribute to the OM. During maturation, some of these compounds are incorporated into the kerogen, although most of these reactive components are rapidly degraded.

Classification of coal

Durand et al. (1977) showed for a series of kerogens isolated from coals that the OM from these coals and Type III kerogen can be classified together in the same type. Furthermore, studies of Combaz and de Matharel (1978) showed that the Eocene-Miocene coal layers from the Mahakam delta, which occur together with kerogen Type III shales, derive from the same precursors. However, investigating humin from very immature coals and shales from the Paris Basin, Huc et al. (1986) showed differences due to sedimentological conditions. Nevertheless, no detectable differences between the structural evolution of coals and shales with dispersed Type III OM at any maturity stages were observed and, therefore, both can be grouped together. However, the depositional environment of massive coal beds differs from deltaic sedimentation sites. Large coal beds were mainly deposited during Carboniferous Age in tropical swamps. In contrast to deltaic sediments, there is no transportation by rivers and, therefore, the terrestrial plant material leads to almost thick and pure organic matter deposits.

Coals are classified as either humic or sapropelic coals (Killops and Killops, 2004). Humic coals are the most abundant class of coal. They are

2.2. Sedimentary macromolecular organic matter - composition, classification and maturation

formed mainly from vascular plant remains, the major organic components deriving from the humification of woody tissue. The main maceral group is vitrinite and the evolution of humic coals is equal to Type III kerogen in Figure 2.5. Examples for humic coals are the Westphalian coals of northern Europe.

Less common than humic coals are sapropelic coals. They form in shallow waters with an oxygen deficit. In contrast to humic coals they do not undergo a peat stage during their maturation but follow the diagenetic pathway of maturation outlined for hydrogen rich kerogens of Type I or II. The sapropelic coals contain varying amounts of allochtonous organic and mineral matter, as the hydrogen rich kerogens do, but the main organic fraction is derived from autochtonous algal remains and, in addition, peat swamp plants. The sapropelic coals are divided into two subgroups, boghead coals (torbanites) and cannel coals. The boghead coals contain a larger amount of algal material and some fungi remains, whereas cannal coals have a higher proportion of spores. In contrast to humic coals, sapropelic coals are able to generate oil during maturation according to their aliphatic rich character.

Coalification

The maturation of coal is here given special attention because it forms a central theme of this thesis. The processes of coal formation can be divided into two main stages: (1) the peatification and (2) the coalification. Furthermore, the coalification process can further be subdivided into a biochemical (early coalification) and a geochemical phase (later coalification) (Killops and Killops, 2004). Peatification and early coalification stages dominated by biological alteration of the OM are equal to the term diagenesis, while the organic matter alteration during later coalification stage is more related to increasing temperature and pressure conditions and is, therefore, equal to catagenesis and metagenesis. The evolution and maturation of coals can be classified by different coal types, first peat, followed by lignite, sub-bituminous coal,

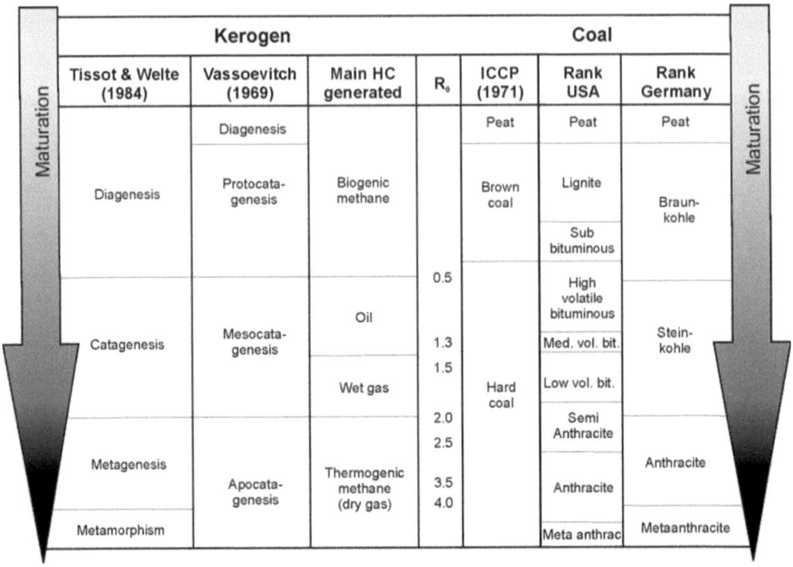

2. Background

	Kerogen				Coal			
Maturation	Tissot & Welte (1984)	Vassoevitch (1969)	Main HC generated	R_o	ICCP (1971)	Rank USA	Rank Germany	Maturation
		Diagenesis			Peat	Peat	Peat	
	Diagenesis	Protocatagenesis	Biogenic methane		Brown coal	Lignite	Braunkohle	
						Sub bituminous		
				0.5		High volatile bituminous		
	Catagenesis	Mesocatagenesis	Oil	1.3		Med. vol. bit.	Steinkohle	
			Wet gas	1.5	Hard coal	Low vol. bit.		
				2.0		Semi Anthracite		
				2.5				
	Metagenesis	Apocatagenesis	Thermogenic methane (dry gas)	3.5		Anthracite	Anthracite	
				4.0				
	Metamorphism					Meta anthrac	Metaanthracite	

Figure 2.12.: *Stages of organic matter evolution and comparisation of different maturity scales for coals and kerogen, the related vitrinite reflectance values and the main generated hydrocarbons (HC), modified after Tissot and Welte (1984b) and Vandenbroucke and Largeau (2007)*

bituminous coal and finally anthracite. Beside these differentiations, there are other types of maturity related classifications for organic matter evolution. The term brown coal summarises the lignite and sub-bituminous coals and is related to diagenetic alteration of the OM. The name hard coal is for the more mature coals related to catagenesis and metagenesis. Figure 2.12 compares different maturity scales for coal and kerogen with regard to the vitrinite reflectance data.

Figure 2.13 shows the position of the different coal types in the van Krevelen diagram. The humic coals follow the pathway of a Type III kerogen. The main changes during diagenesis are related to a decrease in the C/O ratio. Sapropelic coals are more equal to Type II kerogen starting with high

2.2. Sedimentary macromolecular organic matter - composition, classification and maturation

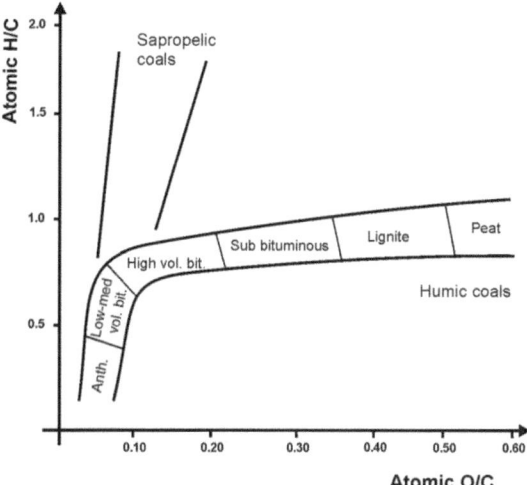

Figure 2.13.: *Maturation stages of coals plotted in a van Krevelen type diagram. Sapropelic coals start with a high H/C atomc ratio due to a high aliphatic content, humic coals start with a high O/C atomic ratio and undergo the coalification stages peat, lignite, sub-bituminous coal and high volatile bituminous coal. Modified, after Killops and Killops (2004)*

H/C and low C/O values and, therefore, the early changes in composition are related to a loss of hydrogen. Due to this, sapropelic coals can act as a source rock for oil generation. For humic coals, the end of diagenesis is correlated with a C/O value of around 0.1.

Peatification is the initial alteration process of plant material alteration when woody material is deposited in swamps. Only 10% of the woody debris contribute to the formation of peat, the remaining organic material is recycled by microbial communities or is lost by leeching or remineralisation. The result of peatification is a brown hydrated gel without internal structures. The initial degradation process during peat formation is the decomposition of polysaccharides by decomposers, microorganisms like bacteria or fungi that use exoenzymes for the destruction of the carbohydrate polymer (Hatcher and Clifford, 1997). During peat formation, all of the cellulose, that

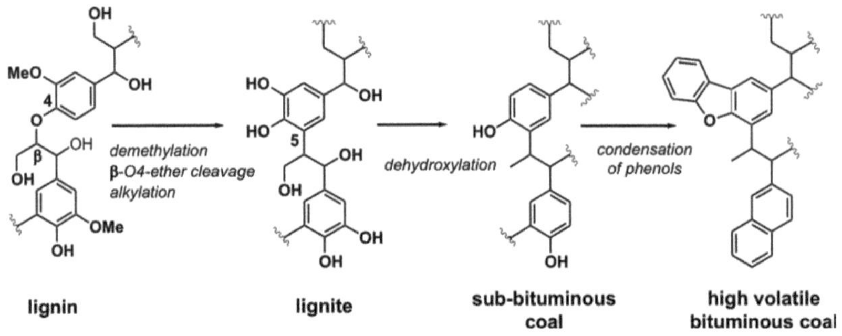

2. Background

Figure 2.14.: *Reactions transformig lignin to lignite, sub-bituminous coal and high volatile bituminous coal, modified after Hatcher and Clifford (1997) and Killops and Killops (2004).*

is the major component of wood, is converted into glucose and is degraded (Hatcher et al., 1982; Stout et al., 1988). The main contributor to peat is the more stable biomacromolecule lignin as well as lipids from leaves, spores and pollen. These are minor compounds in higher plants but they are accumulated during peatification due to their resistance towards degradation. Peat in general has a high water content of up to 95%.

Lignite formation during the early coalification stage is dominated by microbial alteration of lignin by a number of enzymatic reactions (Hatcher, 1990). The content of lignin decreases with increasing maturation, and in peats, it can account for about 35%, decreasing down to less than 10% in sub-bituminious coals. Early coalification is dominated by condensation reactions leading to the formation of a complex geopolymer, being accompanied by the loss of oxygen containing groups. The main reaction is the cleavage of aryl ether bonds by hydrolysis of methoxy groups and the β-O-4 ether cleavage (Bates and Hatcher, 1989; Hatcher et al., 1989). These reactions lead to the formation of phenols in the lignite structure (Figure 2.14). Carbocations generated in the β-O-4 cleavage reaction undergo alkylation reactions e.g. alkylation of the C5 position in Figure 2.14 leading to the macromolecular structure of lignite. The aromaticity of the coalified wood does not increase during this stage of coalification. Ongoing matu-

2.2. Sedimentary macromolecular organic matter - composition, classification and maturation

ration of the lignitic wood debris includes significant structural changes. These changes are most likely the primary reasons for the visible changes in physical morphology. The vitrinite macerals become more visually and microscopically homogenous. The plant material cells become deformed due to increased pressure to form a homogeneous mass. Chemically, the main degradation reaction during sub-bituminous coal formation is the loss of oxygen, mostly by the dehydroxylation of catechols (dihydroxy phenols) resulting in an increasing amount of monohydroxy units (Hatcher, 1990; Hatcher and Clifford, 1997). The main compound units in lignite, the catechols, are transformed to phenols in sub-bituminous coals. In addition to this, the dehydroxylation of side chain hydroxy groups also occurs.

The transformation to high volatile bituminous coals is accompanied with an additional loss of oxygen, but the aryl-O content does not change significantly. The main reaction is the condensation of phenols to aryl ethers and benzofuran like structures (Figure 2.14) (Hatcher et al., 1992). Also monoaromatic structures condense during this maturation phase forming naphthalenes and fluorenes, being visible in the pyrolysates of high volatile bituminous coals. The degree of aromaticity increases at this rank of maturation (Hatcher, 1988).

It can be seen from the maturation pathway for humic coals in the van Krevelen diagram (Figure 2.13) that the decrease of the O/C values is terminated at this maturation stage. Subsequent alteration of the organic matter is dominated by hydrogen loss, indicating further aromatic condensation reactions. Increasing pressure (and temperature) results in the formation of more and more tabular polyaromatic layers, finally causing the transformation to anthracite.

2.3. Methods to elucidate the structural information of macromolecular organic matter

2.3.1. The structural composition of macromolecular organic matter as a source of information for organic geochemists

Investigation of the structural composition of sedimentary organic matter is one of the main sources of information for organic geochemists. Revealing the compounds that had formed the buried organic matter opens a fascinating view on the evolution of our planet. These molecular witnesses of the past provide information on the sedimentary environment, climatic condition during sedimentation, but also on the maturity of the organic matter and, therefore, on its geological history. Diverse vegetations have left their marks as well as microbial communities forming populations inhabiting the sediments throughout millions of years. Revealing the structural composition of sediments is, therefore, like reading in a history book that contains a vast number of information carved in stone in an encrypted language. Organic geochemists use various analytical tools and chemical methods as keys to decode this information. Those methods can be principally divided into: (1) non-destructive "simple" measurements, spectroscopic methods that directly use the adsorption, emission, reflection or scattering of electromagnetic waves by the sample material (Chapter 2.3.2), (2) non-selective destruction and fragmentation of the macromolecules followed by fragment analysis (Chapter 2.3.3) and (3) the selective and controlled degradation of individual parts of the geopolymer by chemical reactions prior to fragment analysis (Chapter 2.3.4).

Whenever geochemists collect data from a geological sample set, they are confronted with the fact that all the response signals to their measurement techniques are detected as a superposition of various signals of the highly complex material. Therefore, the identification of the received signals has to be done very carefully. Interpretation of the extracted data set is also

2.3. Methods to elucidate the structural information of macromolecular organic matter

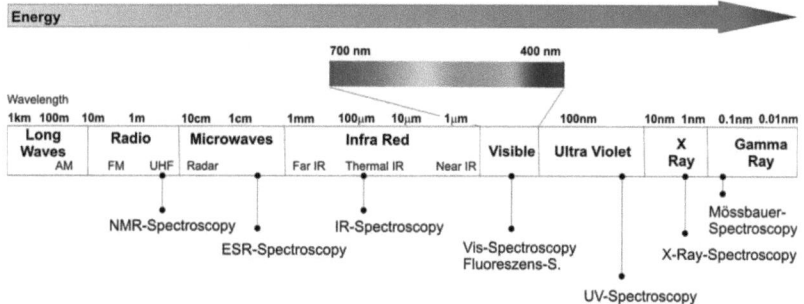

Figure 2.15.: *The electromagnetic spectrum from low energy (long wavelength, AM) to high energy content (short wavelength, gamma ray) can be used for analytical techniques.*

challenging due to the complex mechanisms that once caused the structure found in the present day sediments and, therefore, has also to be carried out with care.

2.3.2. Spectroscopic methods

A simple way of gaining information concerning molecular organic matter is by the use of electromagnetic waves. Molecules can interact with well defined energy packages (quanta) by absorbing a specific energy amount. The absorbed energy is normally emitted subsequently as a well defined energy quantum. Thus, the whole electromagnetic spectrum can interact with different energy stages of the molecule. Spectroscopic methods use these interactions to receive information from absorption or emission of this energy as well as in some cases scattering or reflection of specific wavelengths. Figure 2.15 gives an overview of the electromagnetic spectrum and the measurement techniques that use specific ranges of that spectrum.

Mössbauer spectroscopy uses the high frequent gamma rays and, therefore, interacts with atomic nuclei (Mössbauer, 1962). The absorption and emission of a gamma quantum by an atomic nucleus is influenced by the surrounding electrons. It is used to separate iron(II) from iron(III) and to

2. Background

detect some other metals in rocks (e.g. tin, antimony and tellurium), but its application is limited to inorganic chemistry (Gütlich, 1970). Therefore, the Mössbauer spectroscopy is not discussed further in this thesis.

X-rays and XANES spectroscopy

X-ray spectroscopic methods are used for example in computed tomography (CT) to receive an insight into whole rocks (Jacobs and Cnudde, 2009) or to receive information about element specific oxidation state or electron structures.

X-ray quanta can only excite the nucleus near electrons and can promote them into the next free level or they are emitted. When the energy of the x-ray is in resonance with the energy of a specific electronic level, an electron of this level absorbs the energy and leaves this level. The absorption is visible by an edge in the spectrum. This edge is called K-edge if the absorbing electron is located in a K level or L-edge if the electron is located in the L level. The resulting hole in the nucleus near electron level is than refilled by an electron from a more distant level. The resulting energy of this refilling process can be released not only by a photon but also by an electron emitted from an outer level (Auger electron). These processes can be used to measure the X-ray absorption energy from an atom in the X-ray absorption near-edge structure (XANES) spectroscopy (Huffman et al., 1991, 1995; Charrié-Duhaut et al., 2000). This technique is element specific and sensitive to the electron structure, to the nucleus-electron attraction (depending on the amount of electrons and nuclei and the oxidation stage) as well as to the local symmetry of the absorbing site. Therefore, XANES spectroscopy is used in organic geochemistry e.g. for the estimation of the sulfur content such as thiophenic, alkyl or aryl sulfur (Brown et al., 1992) or oxidised sulfur forms such as sulfoxides, sulfones and sulfates (Sarret et al., 1999) in bitumen samples.

2.3. Methods to elucidate the structural information of macromolecular organic matter

Visible and ultra violet light

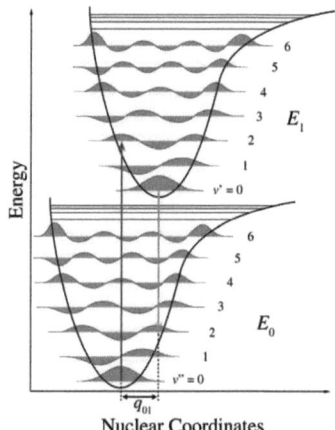

Figure 2.16.: *After absorption of energy (blue arrow) the electron is promoted from the ground state (E_0) to the excited level (E_1). Energy emission (green arrow) when returning to the ground state is of less energy (longer wavelength) due to the energy loss by vibration (ν: vibration modes) (Franck-Condon principle). Picture taken from* http://en.wikipedia.org/wiki/Franck_condon.

The wavelength of ultra violet and visible light provides energy that interacts with the electrons from the outermost orbitals in molecules (HOMO = Highest Occupied Molecular Orbital). Absorption of the energy quanta from this range of the electromagnetic spectrum promotes the molecules to an excited stage where some electrons of the molecule are promoted to a higher energetic level ($E_0 \rightarrow E_1$, Figure 2.16). The absorption can be measured directly and the absorbed wavelength is characteristic for a specific electron situation in the molecule. In aqueous solutions of humic acids and fulvic acids the E_4/E_6 ratio (ratio of the extinction at 465 and 665 nm) can be determined and used as a characterisation of the degree of aromatisation (Chen et al., 1977; Christl et al., 2000; Khanna et al., 2008).

In aromatic molecules, the energy absorption in the range of UV light wavelength is due to delocalised π-electrons. When excited electrons return to the ground state, the energy is emitted again, with a little shift to longer wavelength (q_{01} in Figure 2.16) due to transition from vibration modes $\nu_2 \rightarrow \nu_0$ in E_1 and E_0 that reduces the emitted energy during the transition from the excited stage E_1 back to the ground stage E_0 in Figure 2.16 (Franck-Condon principle). Therefore, the emitted light is of less energetic visible light (large wavelength), an effect known from fluorescence of aromatic

2. Background

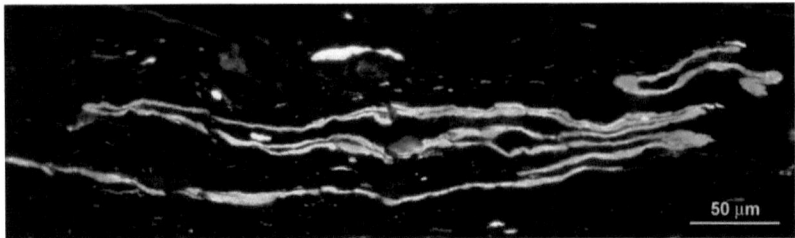

Figure 2.17.: *The maceral liptinite shows intense fluorescence in fluorescence microscopy. Picture with permission of Andreas Fuhrmann.*

molecules. When the concentration of aromatic molecules in an environment (e.g. in a solution or in a maceral) is very high, the fluorescence disappears due to fluorescence quenching. Therefore, only macerals with a low concentration of aromatic molecules show intense fluorescence. This can be used to identify liptinite rich organic matter, bearing a potential for oil generation, by fluorescence microscopy (Figure 2.17).

In the maceral vitrinite, the dominating aromatic structures become more and more flat during maturation due to aromatisation reactions leading to an increase of aromatic structures. White light is reflected by these flat aromatic structures and the amount of reflectance increases with increasing aromatisation. Therefore, the vitrinite reflectance (R_0 measured in %) is used to determine the thermal maturation of organic matter (containing vitrinite macerals) in sediments.

Infrared (IR) spectroscopy

Infrared light interacts with the rotation and vibration modes of molecules, atoms and atom groups in molecules. It is divided into near-infrared (NIR) comprising wave numbers from 12500 cm^{-1} to 4000 cm^{-1} (λ: 0.8 - 2.5 μm), the mid-infrared (MIR) from 4000 cm^{-1} to 400 cm^{-1} (λ: 2.5-25 μm) and the far-infrared (FIR) from 400 cm^{-1} to 10 cm^{-1} (λ: 25 - 1000 μm). FIR waves induce rotation of complete molecules and NIR is used to excite overtone or

2.3. Methods to elucidate the structural information of macromolecular organic matter

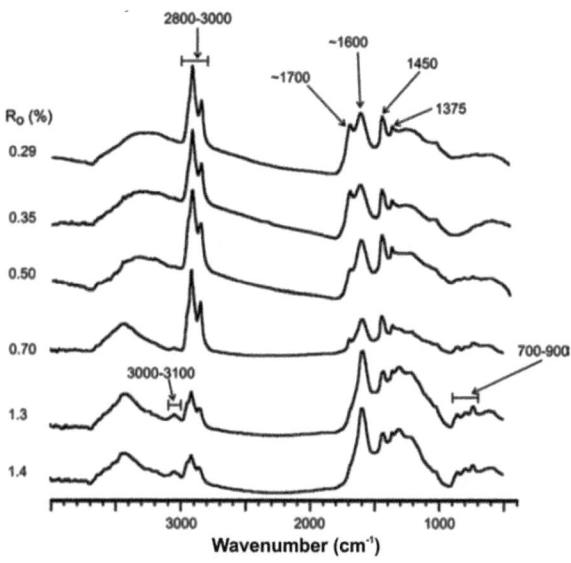

Figure 2.18.: *IR spectra of kerogens of different thermal maturation, modified from Lis et al. (2005). CO peaks (≈ 1700 cm^1) and aliphatic C-H peaks (2800 - 3000 cm^1) decrease, aromatic C=C peaks (1600 cm^1) and aromatic C-H peaks (3000 - 3100 cm^1) increase with ongoing maturation, reflecting an increase of aromaticity.*

harmonic vibrations. The MIR is the "normal" IR that excites the vibration of atoms and atom groups (CH_3, OH, NH_2 etc.) inside molecules such as stretching, scissoring, rocking, wagging and twisting. The resonance frequency of each vibration of an atom or atom group depends on the bond strength and the weight of the atoms (large and heavy atoms vibrate slower and at lower frequency than small and light atoms). Therefore, MIR can be used to identify functional groups in molecules like for example carbonyl or carboxyl groups (CO stretching: 1670 - 1820 cm^{-1}), hydroxy groups (OH stretching: 3200 - 3700 cm^{-1}) and aromatics (C=C stretching: 1400 - 1600 cm^{-1} and C-H stretching: 3000 - 3100 cm^{-1}) (Lis et al., 2005; Petersen et al., 2008).

2. Background

Since the initial development of Fourier transform infrared spectroscopy (FTIR), the signal to noise ratio and the detection limits have gradually been improved, so that FTIR has become a widely used technique to reveal the structural composition of coals (Painter et al., 1981; Kuehn et al., 1982). Figure 2.18, taken from Lis et al. (2005), shows IR spectra of kerogen samples of different maturity levels between R_0: 0.29% and 1.4% in the wave number range from 4000 - 400 cm^{-1} (MIR). Comparing the absorption bands of these samples, the principal structural changes during maturation can be observed. During diagenesis and the beginning of catagenesis (samples with R_0: 0.29% - 0.70%), the band at 1700 cm^{-1} belonging to carbonyl/carboxyl groups (C=O band), decreases due to the degradation of oxygen containing groups with ongoing maturation. Those samples that have passed the oil window (samples with R_0: 1.3% and 1.4%) show an increase of the absorbtion band at 1600 cm^{-1} belonging to the C=C stretching modes. Additionally a band at 3000 - 3100 cm^{-1} occurs that belongs to aromatic C-H stretching modes, indicating the increase of aromaticity during catagenesis. The loss of aliphatic compounds during the oil window is visible by the decrease of the absoption bands at 2800 - 3000 cm^{-1} that refers to aliphatic C-H stretching modes.

Electron spin resonance (ESR) spectroscopy

The microwave resonance absorption of atoms in molecules can be measured using the electron spin resonance spectroscopy (ESR), also called electron paramagnetic resonance spectroscopy (EPR). This technique can be applied to all elements that contain unpaired electrons and, therefore, have a magnetic moment (paramagnetic elements).

2.3. Methods to elucidate the structural information of macromolecular organic matter

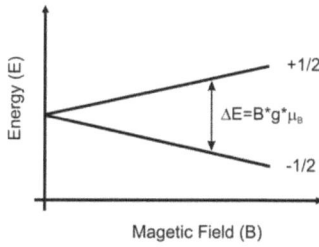

Figure 2.19.: *Separation of electron spin moment by an external strong magnetic field (B) into two spin states (+1/2 and -1/2) with the energy difference ΔE (Zeeman effect, g: g-factor, μ_B: bohr magneton).*

When paramagnetic elements are brought into an external, strong magnetic field, the electrons with spin of $m_S = +1/2$ or $-1/2$ orientate themselves either parallel or anti parallel to the magnetic field. Their energies are no longer equal, the parallel orientated electron is now of a lower energy state than the anti parallel orientated electron (Zeeman effect, Figure 2.19). The energy difference between this states depends on the strength of the external magnetic field (B) as well as the *Bohr magneton* (μ_B, a physical constant of the magnetic moment of electrons) and the *g-factor* (depending on the electron spin moment). By absorbing or emitting an energy quantum of ΔE, the electrons can move between the two energy states. The required magnetic fields to align the spin magnetic moments of electrons are of a size of about 3500 G and the resonance frequencies are in the range of 9 - 10 GHz (Atkins, 1986).

ESR spectroscopy can be used to analyse paramagnetic metals in inorganic chemistry and in organic chemistry to detect and quantify free radicals. This is also the main application in organic geochemistry where free radicals can be found in kerogens or coals during ongoing degradation (Marchand and Conard, 1980). From studies of humic substances, it is known that the number of free radicals increases with progressive humification (Riffaldi and Schnitzer, 1972; Watanabe et al., 2005; Yabuta et al., 2008). The origin of these free radicals is mostly semiquinone where the free radicals are associated with the condensed aromatic network and are stabilized by the aromatic ring system. At high maturity levels, an inflexion point in the free radical concentration occurs because of electron pairing to

2. Background

form new bonds (Horsfield, 1984).

Nuclear magnetic resonance (NMR) spectroscopy

As is the case for electrons, so the atomic nuclei are also characterised by a spin magnetic moment. Due to their different size and mass from electrons, the resonance energy requires longer wavelength in the range of radio irradiation. The required frequencies are in a range of kHz to MHz. As the ESR spectroscopy works only for paramagnetic elements, the nuclear magnetic resonance spectroscopy (NMR) can only be applied to elements with a nucleus of odd spin. Therefore, the nuclei commonly used in NMR spectroscopy are mainly ^{1}H ($S = 1/2$), ^{13}C ($S = 1/2$), ^{15}N ($S = 1/2$), ^{31}P ($S = 1/2$). The nuclei ^{17}O ($S = 5/2$), ^{19}F ($S = 1/2$), ^{29}Si ($S = 1/2$), all alkali metals and some further nuclei are measurable but of minor importance because they are either of minor abundance or relevance.

When the atoms are brought into a strong external magnetic field, the spins orientate their magnetic moment parallel or anti parallel to this magnetic field. As in ESR spectroscopy, the energy difference of these two energy states depends on the strength of the magnetic field (B) and the spin of the nuclei. In the NMR experiment, the sample located in the magnetic field is pulsed with a radio pulse which is orientated 90° to the magnetic field. When the spin frequency is in resonance with the pulse frequency, the spin magnetic moment changes its orientation. Its relaxation after the radio pulse is measured with a receiver coil. The resonance frequency depends on the chemical environment of the nucleus and the coupling of the spin with the spin of its neighbouring nuclei.

In solid samples, there is no signal averaging by thermal motion like in liquid samples and, therefore, slightly different environments will give a different signal and the resolution is lowered. To improve the resolution in the solid-state NMR spectroscopy, the sample is rotated at an angle of ca. 54.74° with respect to the magnetic field named magic spinning angle (MAS)

2.3. Methods to elucidate the structural information of macromolecular organic matter

(Schaefer and Stejskal, 1976). Cross polarisation (CP) of carbon atoms by protons in combination with MAS technique has improved the resolution of the solid-state NMR and the CP/MAS technique has become a standard method for analysis of solids.

The carbon NMR, especially CP/MAS experiments, are commonly used in organic geochemistry for analysing the functional group composition as well as the amount of aromatisation, aliphatic HC content, cross linking and network formation because aliphatic aromatic and carboxylic carbon atoms can easily be separated and quantified in ^{13}C NMR spectra (Wilson and Hatcher, 1988; Bates and Hatcher, 1989; Suggate and Dickinson, 2004).

2.3.3. Pyrolysis

To gain information about the structural units forming the macromolecular matter, pyrolysis is one of the most established approaches in organic geochemistry. The term pyrolysis has been defined as "a chemical degradation reaction that is induced by thermal energy alone" (Ericsson and Lattimer, 1989). When pyrolysis is coupled with gas chromatography using flame ionisation detection (GC-FID) or mass spectrometry detection (GC-MS) the products of thermal cracking reactions in the kerogen can be identified and the detection and quantification of specific fragments are characteristic for the original organic matter composition (Horsfield, 1989).

Pyrolysis can be performed in an open system as well as in a closed system (using sealed vessels while heating the sample). When a sample is heated until thermal cracking reactions start, free radicals will be generated as a result of homolytic bond cracking and e.g. CO_2 expulsion. It can be described as a number of parallel first order reactions characterised by different potential and activation energies (Schenk et al., 1997). The generated radical compounds undergo several consecutive reactions e.g. hydrogen abstraction, elimination or radical recombination. These products can be detected by GC-FID or GC-MS measurement. Pyrolysis products differ

2. Background

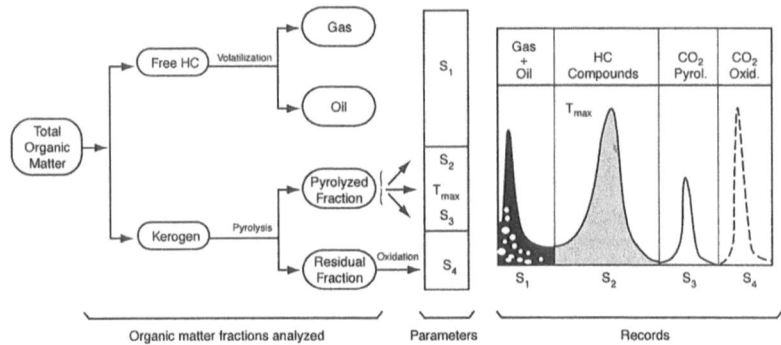

Figure 2.20.: *Scheme of Rock-Eval pyrolysis, showing the different fractions of total organic matter in a rock sample, taken from Lafargue et al. (1998).*

between open and closed system pyrolysis. In an open system, the initial cracking products are transported immediately from the pyrolysis unit and, for example, alkyl radicals will react either in radical recombination or hydrogen abstraction from another molecule to give an alkane or in hydrogen elimination to give an alkene. Therefore, alkane/alkene doublets always appear in the chromatogram. In a closed system, radicals remain within the pyrolysis cell and recombination is the preferred consecutive reaction. Therefore, mainly alkanes and only very low amounts of alkenes are visible in the chromatogram.

Using pyrolysis, the geochemical cracking reactions that lead to the generation of oil within a sedimentary basin and the maturation of OM can be simulated very effectively when applying different heating rates in the pyrolysis experiments (Horsfield et al., 1989). The kinetic parameters (activation energy and frequency factor) of kerogen can be determined with this method (Burnham et al., 1987; Mahlstedt et al., 2008).

A simple and powerful application for the evaluation of sedimentary rock samples is Rock-Eval pyrolysis (Espitalié et al., 1985; Lafargue et al., 1998; Béhar et al., 2001). In this technique, the initial sediment sample is heated

2.3. Methods to elucidate the structural information of macromolecular organic matter

first up to 300°C to mobilise all free (unbound) compounds within the rock sample. The amount of mobilised compounds is then analysed by an FID to provide a bulk signal, the S1 peak (Figure 2.20). This is a rapid estimation of the C_{6+} hydrocarbon content of a rock sample. In a further step, the rock sample is heated from 300°C up to 550 or 600°C (depending on the instrument model, or even 850°C with the Rock-Eval 6 instrument) to perform pyrolysis of organic macromolecules. The products of thermal cracking are analysed by FID to give the bulk parameter for the total releasable amount of hydrocarbons from the rock samples. This parameter is known as the S2 peak. The released CO_2 during heating up to 390°C is trapped in a molecular sieve and is measured afterwards by a thermal conductivity detector (TCD) to analyse the amount of CO_2 released by the thermal cracking reactions and to obtain a bulk parameter for the oxygen content of a rock sample, named the S3 peak. CO_2 trapping is terminated at 390°C because at higher pyrolysis temperatures, the produced CO_2 may also derive from inorganic carbonates in the sediment. The remaining residue can further be oxidised with oxygen by burning the sample at 850°C under a flow of air and the obtained CO_2 can be measured in the same way as done for the S3 to give the S4 peak that provides information about the content of oxidisable carbon within the residue and to evaluate the amount of residual organic carbon (ROC).

These bulk parameters can be used to calculate important parameters for the characterisation of sedimentary rocks. These are the Bitumen Index (BI=S1/TOC) showing the amount of free hydrocarbons in the organic fraction, the Hydrogen Index (HI=S2/TOC) and the Oxygen Index (OI=S3/TOC) indicating the amount of hydrogen and oxygen in the kerogen, the Quality Index (QI=(S1+S2)/TOC) giving insight into the total oil generation potential of a kerogen and the Production Index (PI=S1/(S1+S2)) indicating the degree of transformation from bound precursor substituents into the free hydrocarbons. Furthermore, the T_{max} value can be determined by Rock-Eval pyrolysis. This is the temperature where a sample produces the highest

2. Background

rates of pyrolysate generation (temperature of the maximum of the S2 peak). The T_{max} value provides a maturity related parameter. A low mature sample produces high amounts of hydrocarbons at low pyrolysis temperature (low T_{max}) and a high mature sample produces the most hydrocarbons at higher pyrolysis temperature (high T_{max}).

2.3.4. Chemical degradation

For a deeper understanding of geochemical processes that lead to the formation of macromolecular OM such as coals or kerogen in source rocks as well as for a general understanding of the geological carbon cycle and its coupling to the biosphere, often a more detailed view into organic matter composition is necessary. To understand changes within the sediments with ongoing maturation of organic matter precise information about the chemical structure and the bond situation is very important. The use of selective chemical cleavage reactions coupled with analytical measurement of the reaction products provide a powerful tool to obtain the required information. Investigation of the cleavage products give information about the types of lipids bound to the kerogen and the nature of the bonds linking these compounds to the macromolecular matrix (Rullkötter and Michaelis, 1990; Richnow et al., 1992; Schaeffer et al., 1995; Schaeffer-Reiss et al., 1998).

Chemical reagents can attack individual types of bonds selectively and release, therefore, only those compounds that were linked by a given specific bond type. Heteroatoms appearing within sedimentary organic matter are in general oxygen (acids, hydroxides, esters or ethers), in some cases sulfur (forming S-S or C-S bonds) and, in very young sediments, nitrogen (as amines or amides). Several individual cleavage reactions are available to selectively attack these types of chemical bonds. These reactions can be performed in a single procedure or they can be applied in combination to achieve a stepwise degradation of the sample material (Figure 2.21) (Richnow et al., 1992; Schaeffer et al., 1995; Höld et al., 1998; Schaeffer-Reiss

2.3. Methods to elucidate the structural information of macromolecular organic matter

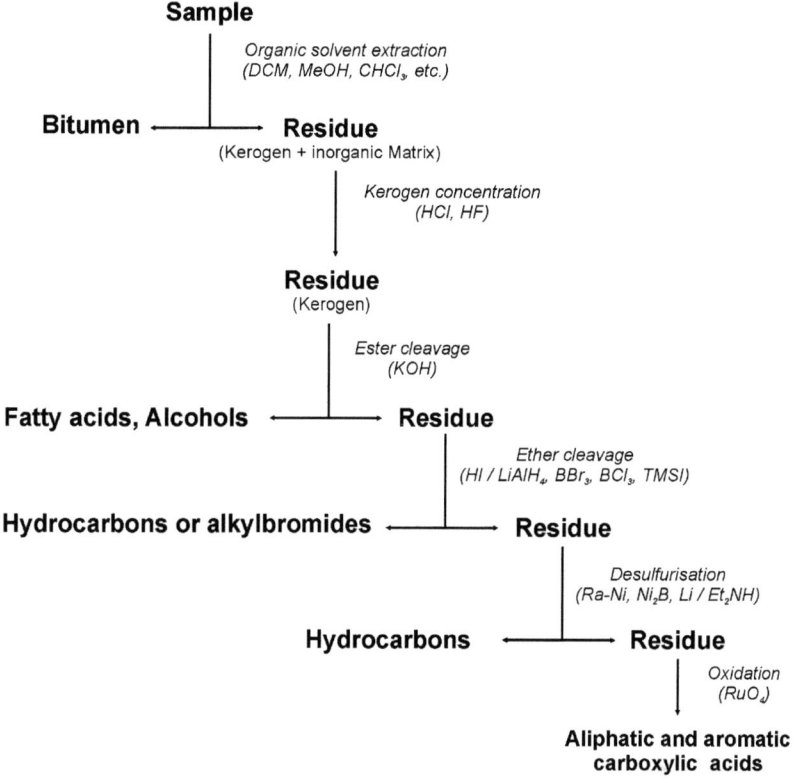

Figure 2.21.: *Scheme of a stepwise chemical degradation procedure.*

et al., 1998).

2. Background

Ester and amide cleavage (hydrolysis)

The most labile oxygen-containing bond within macromolecular OM is the ester bond. In immature sediments, also nitrogen linkages are observed, forming amides. Amide and ester bonds can be hydrolysed in a catalytic process either in acidic (amides and esters) or in alkaline environment (only esters). The acid catalysed ester hydrolysis is a reversible reaction. The free acids can react with the alcohols to reform esters. In the alkaline ester cleavage reaction, the liberated acids become deactivated by forming the corresponding salt (Exner, 1988; Glombitza et al., 2009b).

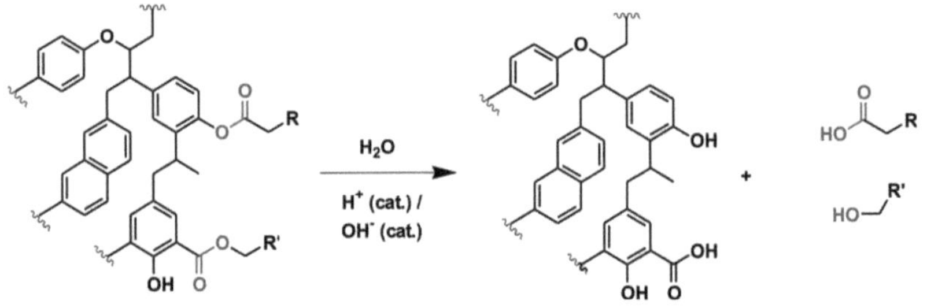

Figure 2.22.: *Scheme of ester cleavage in coals, liberating terminally bound organic acids and alcohols. The catalysts for this reaction can be either protons (acid hydrolysis) or hydroxy ions (alkaline hydrolysis, saponification).*

Ester cleavage results in free carboxylic acids and alcohols (Figure 2.22), the cleavage of amides results in acids and amines. When applying this reaction to kerogen, the terminal linked acids, alcohols and amines are liberated and can be extracted using an organic solvent. The non-terminal conjugated bonding partner will remain as a part of the kerogen residue. For alkaline hydrolysis of esters (saponification), the most common base used is potassium hydroxide disssolved in a polar solvent, e.g. methanol (Höld et al., 1998; Putschew et al., 1998; Schaeffer-Reiss et al., 1998). The reaction requires heat and a reaction time of several hours to complete. Cleavage products obtained from macromolecular OM are fatty acids, alcohols, aro-

2.3. Methods to elucidate the structural information of macromolecular organic matter

matic acids and phenols as well as larger fragments of the macromolecules liberated by ester bond cleavage such as humic and fulvic acids.

Ether cleavage

The cleavage of ether bonds in kerogen can be achieved by treatment with boron trichloride (BCl_3) (Chappe et al., 1982; Richnow et al., 1992), boron tribromide (BBr_3) (Schwarzbauer et al., 2003), hydrogen iodide (HI) (Schaeffer-Reiss et al., 1998; Putschew et al., 1998; Höld et al., 1998), trifluoroacetic acid (Bhatt and Kulkarni, 1983), boron trifluoride (Narayanan and Lyer, 1965) or trimethylsilyiodide (TMSI) (Michaelis and Richnow, 1989). This reaction will also cleave esters if hydrolysis is not performed prior to the ether cleavage. The products of the ether cleavage reaction are mono- or difunctionalised halogen derivatives in case of mono- or diethers. Applying a further reduction using lithium aluminum hydride ($LiAlH_4$) results in the formation of the corresponding alkanes. Labelling can be achieved using a deuterated reagent ($LiAlD_4$) (Figure 2.23). In this case, the former oxygen bearing position can be identified by mass spectroscopy and with this the position of the former ether bond.

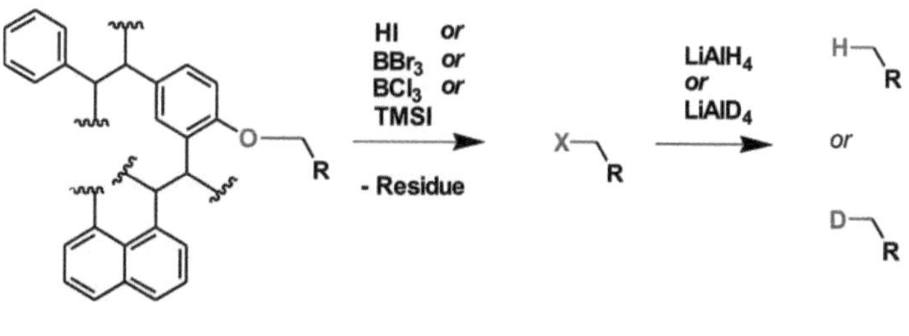

Figure 2.23.: *Scheme of ether cleavage in coals, liberating terminal bound alcohols. Labelling of the former bond position can be achieved using $LiAlD_4$. X = halogen (Cl, Br, I).*

2. Background

Sulphur cleavage

Carbon-sulphur and sulphur-sulphur bonds can be cleaved using a radical reduction reaction with hydrogen and reactive metals such as lithium or nickel (Figure 2.24). Nickel is used as the reactive Raney Nickel, an alloy of nickel and aluminium which is used as a catalyst for hydrogenation (Sinninghe Damsté et al., 1989b; Sinninghe Damsté and De Leeuw, 1990). A further nickel reagent for desulphurisation is nickel boride (Ni_2B) which is generated in-situ from $NiCl_2$ and $NaBH_4$. The required hydrogen being part of the newly formed C-H bonds, is derived either from the $NaBH_4$ or from the used protic solvent (H_2O or ROH) (Back et al., 1992). Due to the in-situ generation of "Ni_2B", less activation problems occur compared with the use of Raney Nickel resulting in higher product yields (Back et al., 1992; Schouten et al., 1993; Sinninghe Damsté et al., 1999).

An alternative approach for the cleavage of sulphur bonds is the use of lithium. When lithium is dissolved in a dry amine (e.g. NH_3, $EtNH_2$ or Et_2NH, etc.), dissolved electrons form that act as a reducing agent. The use of $Li/EtND_2$ offers the possibility to label the former linkage position (Benkeser et al., 1955; Hofmann et al., 1992; Schaeffer-Reiss et al., 1998).

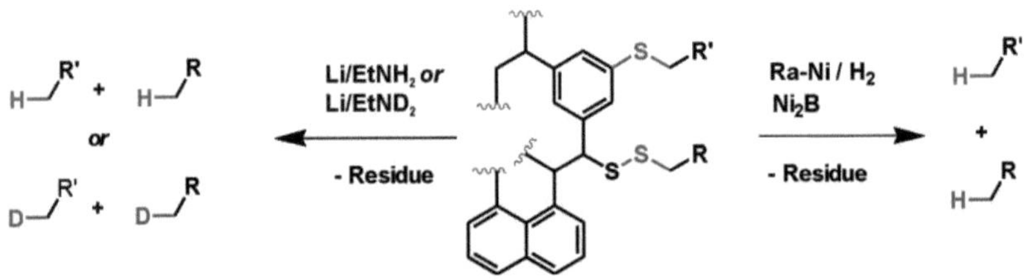

Figure 2.24.: *Scheme of C-S and S-S cleavage in coals, liberating terminal sulphur bound compounds. Labeling of the former bond position can be achieved by using $Li/EtND_2$.*

2.3. Methods to elucidate the structural information of macromolecular organic matter

Oxidation of residual organic matter

To break the macromolecular structure into smaller fragments, strong oxidation reactions can be applied. As a reagent for oxidation, ruthenium tetroxide was introduced in the early fifties by Djerassi and Engle (1953). Applying this oxidation agent to macromolecular organic matter results, for example, in the oxidation of double bonds and the formation of carboxylic acids of aromatic and aliphatic fragments deriving from the degradation of the kerogen matrix and CO_2 (Figure 2.25) (Carlsen et al., 1981; Stock and Tse, 1983; Stock and Wang, 1986; Schaeffer-Reiss et al., 1998).

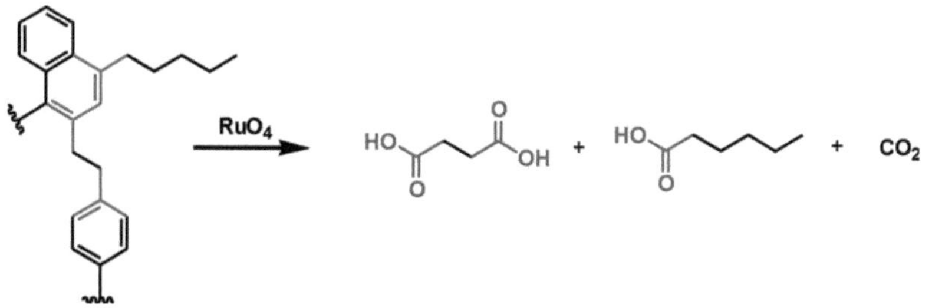

Figure 2.25.: *Scheme of the oxidation of coal with RuO_4 resulting in aliphatic carboxylic acids.*

3. Sample Material And Study Area

3.1. Sample material

3.1.1. The coal mine sample set

Samples for this study were gathered from coal mines and drilled cores of both the North and the South Islands of New Zealand (Figure 3.1) in cooperation with the Institute of Geological & Nuclear Science New Zealand (GNS Science). 16 coal samples were selected from different coal mines that cover a maturity rank from 0 to 11.8 on the scale of Suggate (2000) and vitrinite reflectance of 0.27% to 0.8% (Table 3.1). The coal mine samples cover a nearly continuous range from peat to sub-bituminous coals and are of diagenetic to catagenetic coalification levels. Thus, they present a sample set of coals of low to moderate coalification stage. The coal samples were provided by GNS Science in 2002. They have been stored for several years at room temperature at GNS Science and scince 2002 at the GFZ Potsdam.

The coal mines are located in five different sedimentary basins: Northland Basin, Waikato Basin, Taranaki Basin, West Coast Basin and Eastern Southland Basin (Table 3.1, Figure 3.1). The most immature samples are two samples from the Northland Basin being in peat to lignite maturation stage. The two coal samples of the Eastern Southland Basin are in the lignite maturity range. Samples of the Waikato Basin (six samples) are all sub-bituminous coals and samples from the West Coast Basin (four samples) and the Taranaki Basin (two samples) represent high volatile bituminous coals.

3. Sample Material And Study Area

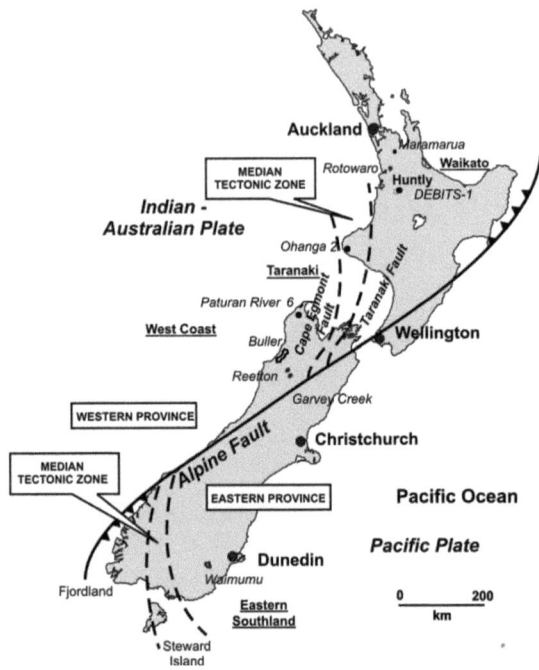

Figure 3.1.: *Map of New Zealand, showing the location of the samples chosen for this study, the sedimentary basins, the coalfields, the three basement zones: Eastern Province, Western Province and Median Tectonic Zone (dotted line: range of the basement zones) and the location of the DEBITS-1 well.*

The selected coals of this sample series present a homogeneous maturation sequence (Vu, 2008) from Late Cretaceous to Pleistocene Age and, therefore, provide excellent sample material for the investigation of maturation related transformations of the organic matter in low mature coals during coalification. Furthermore, they represent a suitable sample set for the assessment of the generation of low molecular weight organic compounds from the kerogen during early maturation processes forming a potential feedstock for deep microbial life.

Table 3.2 shows further analytical data of the coal sample series. These

3.1. Sample material

Sample	Basin	Coalfield	Formation	Age	R_0 [%]	S_R
G001988	Northland	Sweetwater	unknown	Pleistocene	0.27	0
G001986					0.27	0.6
G001976	Eastern Southland	Kapuka	Gore Lignite	Oligocene-Miocene	0.29	1.6
G001978		Waimumu	Measures		0.28	3
G001983		Maramarura			0.41	4.7
G001982		Rotowaro			0.4	5.6
G001984	Waikato	Huntly	Waikato Coal	Eocene-Oligocene	0.45	6.1
G001981		Rotowaro	Measures		0.45	6.6
G001992		Rotowaro			0.49	6.9
G001980		Rotowaro			0.44	7
G001995		Reefton			0.52	7.4
G001996	West Coast	Reefton	Brunner Coal	Eocene	0.52	8.3
G001993		Garvey Creek	Measures		0.76	10.1
G001989		Buller			0.69	11.6
G001994	Taranaki	unknown	Rakopi	Late Cretaceous	0.61	9.5
G001991			Mangahewa	Eocene	0.8	11.8

Table 3.1.: *Selected samples from different coal mines of the North and the South Island of New Zealand, including (GFZ) sample number, sample location, age and maturity rank data, provided by Richard Sykes, GNS Science (New Zealand). S_R=Suggate rank (Suggate, 2000).*

are elemental contents of hydrogen, carbon, nitrogen, sulfur and oxygen in % obtained on a dry ash free (daf) basis, calorific value and volatile matter on a dry, mineral matter and sulfur free (dmmsf) basis and O/C and H/C atomic ratios from dry, mineral matter free (dmmf) basis. Furthermore, in this table, the Rock Eval parameters S1, S2, S3 (from CO_2), T_{max}, TOC, HI (hydrogen index, S2/TOC), OI (oxygen index, S3/TOC), BI (bitumen index, S1/TOC) and QI (quality index, (S1+S2)/TOC) are presented. All data in Tables 3.1 and 3.2 were provided by R. Sykes (GNS Science, New Zealand).

3. Sample Material And Study Area

Sample	C	H	N	S	O	CV [Btu/lb] (dmmsf)	VM [%]	O/C	H/C (mmf)	S1	S2 [mg/g Sed]	S3	T_{max} [°C]	TOC [%]	HI	OI [mg/g TOC]	BI	QI
		[%] (daf)																
G001988	63.34	4.86	0.72	1.18	29.89	10557	58.73	0.343	0.901	10.07	73.50	81.36	370	45.10	163	180	22	185
G001986	64.88	4.77	0.75	1.23	28.37	11076	57.64	0.318	0.863	8.90	82.53	69.86	380	44.86	184	156	20	204
G001976	66.14	4.68	0.59	1.25	27.34	11389	54.58	0.304	0.838	7.44	93.62	61.14	387	50.43	186	121	15	200
G001978	67.83	4.70	0.72	0.76	26.00	11796	51.29	0.281	0.820	2.21	83.66	46.85	398	52.44	160	89	4	164
G001983	71.17	4.85	1.48	0.23	22.27	12434	48.85	0.230	0.809	3.07	93.12	40.20	411	53.82	174	75	6	179
G001982	72.25	4.92	1.17	0.39	21.28	12803	47.90	0.218	0.811	1.26	102.27	25.71	419	54.81	188	47	2	189
G001984	73.46	4.95	1.23	0.28	20.09	12800	45.80	0.202	0.803	0.76	113.56	24.20	416	63.33	180	38	1	181
G001981	73.20	4.78	1.12	0.28	20.61	13123	46.17	0.208	0.778	0.72	103.93	28.67	420	60.75	172	47	1	172
G001992	n.d.	n.d.	n.d.	0.26	n.d.	13254	46.05	n.d.	n.d.	1.51	109.74	29.65	416	60.16	184	49	3	185
G001980	74.79	5.05	1.08	0.30	18.78	13246	45.60	0.179	0.795	0.40	117.68	15.13	418	61.16	194	25	1	193
G001995	75.77	4.95	1.10	0.64	17.54	13324	44.51	0.172	0.782	0.68	128.21	13.45	420	67.76	191	20	1	190
G001996	76.82	5.39	1.27	1.13	15.39	13787	44.86	0.143	0.829	0.98	165.39	8.75	419	67.85	245	13	1	245
G001993	81.51	5.32	1.08	1.56	10.53	14442	43.20	0.095	0.779	2.78	194.28	6.09	430	77.73	252	8	4	254
G001989	77.53	5.24	1.28	4.21	11.74	14962	41.03	0.111	0.807	6.52	182.67	5.47	429	73.33	251	7	9	258
G001994	n.d.	n.d.	n.d.	0.51	n.d.	14263	44.31	0.120	0.843	2.84	171.54	15.11	426	63.99	270	24	4	273
G001991	n.d.	n.d.	n.d.	1.20	n.d.	15046	40.68	0.068	0.823	9.40	197.53	3.58	434	73.83	269	5	13	280

Table 3.2.: *Geochemical data for the coal sample series selected for the current study, provided by R. Sykes, GNS Science (New Zealand). CV=calorific value; VM=volatile matter; n.d.=no data; S1 and S2 given by mg HC/g Sed; S3 given by mg CO_2/g Sed; HI, BI, QI are given by mg HC/g TOC; OI is given by mg CO_2/g TOC.*

3.1. Sample material

The data from table 3.2 provide information about the thermal maturation of the sample set. With ongoing maturation, the percental amount of carbon atoms increases whereas the percental amout of oxygen atoms decreases. This results in a slight decreasing H/C and a strong decreasing O/C atomic ratio. The strong deacrease of the O/C atomic ratio is in accordance with the thermal maturation pathway of a Type III kerogen when plotted in a van Krevelen type diagram (*cf.* Figure 2.13). With ongoing maturation of the samples, the TOC content increases, reflecting the increasing carbon atom percentage by losing O and H atoms. The strong decrease of the S3 value from Rock Eval analysis also indicates the loss of oxygen bearing functional groups during thermal maturation. The temperature of the maximum amount of hydrocarbon generation (S2 peak) upon heating (T_{max}) is shifted to higher values with ongoing maturation, pointing to the generation of free hydrocarbons by losing the weakest kerogen-bound compounds first.

Overall, these data indicate that the chosen sample set provides a good maturation series to investigate the loss of oxygen bearing compounds from the kerogen and, therefore, the potential supply of substrates for microbial metabolic processes of a buried microbial ecosystem.

3.1.2. The DEBITS-1 well

A second sample set was taken from the 148 m deep DEBITS-1 well which was drilled in February/March 2004 on an open ground near the small village of Huntly located on the North Island of New Zealand (37°29'26.2351"S, 175°10'51.0115"E; Figure 3.1) at Ohinewai in the Waikare Coalfield. The drilled sediments were located in the Waikato Coal Area. This drilling campaign was exclusively dedicated to deep biosphere research within the scope of the international DEBITS project being conducted by biogeochemists and geologists from GFZ Potsdam (Germany), geologists from GNS Sciences (New Zealand) and microbiologists from the University of Cardiff (UK). Precautions

3. Sample Material And Study Area

were taken to control potential contamination of the core material by surface microorganisms by adding fluorescent microbeads (being detectable under a microscope) to the drill mud(Kallmeyer et al., 2006).

The DEBITS-1 well penetrated a complex succession of interbedded organic carbon-rich layers and coarser grained mudstones, siltstones and sandstones. At a depth of about 76 m, the core intersected an unconformity. Below the unconformity moderate indurated sediments comprise sandstones, mudstones (siltsones and claystones) and imbedded coal layers of sub-bituminous rank forming the Te Kuiti Group. Above this unconformity, Late Miocene to Late Pleistocene sediments were deposited, forming the Tauranga Group. This succession comprises poorly consolidated gravels, sands, muds and imbedded lignite layers. The sediments located below the unconformity were previously buried to an estimated depth of about 2000 m and at approximated temperature of up to 75°C, prior to a Middle to Late Miocene uplift. Thus, the coals of the Te Kuiti Group are of higher maturity (R_0 ~0.39%) than the lignites of the overlying Tauranga Group (R_0 ~0.29%) (Vieth et al., 2008) (Figure 3.2). The coals and coaly mudstones of the Tauranga Group have moisture contents of about 54.9-65.6% and the bitumen content ranges between 7.7-53.4 mg HC/g TOC. In contrast, the moisture and bitumen content of the coaly layers of the Te Kuiti Group are much lower, 23.2-27.2% and 0.9-3.6 mg HC/g TOC. The Tauranga Group sediments contain five groundwater aquifers at 3.0-17.0, 23.8-25.5, 29.3-30.9, 34.6-39.3 and 66.7-67.0 m below surface (Vieth et al., 2008). Although the groundwater measurements were contaminated by drill mud, measured temperature and pH data were mainly within the range 16.5-18.1°C and 6.4-8.3, respectively.

For this study, 16 samples from different intervals of the DEBITS-1 well were selected including 13 lignite, mudstone and sandstone samples of the Tauranga Group and three coals of Te Kuiti Group. The exact location and lithology of the sample set is given in Table 3.3.

The samples (except G004541 and G004544) were chosen to provide a

3.1. Sample material

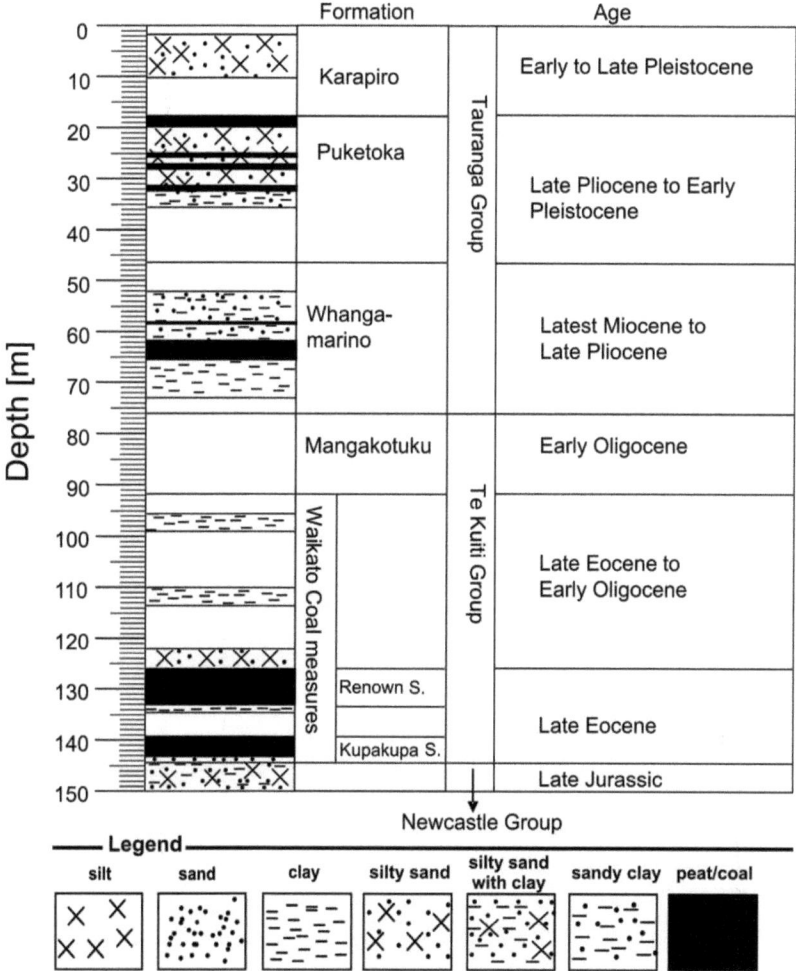

Figure 3.2.: *Stratigraphy of the DEBITS-1 well including formations, age and stratigraphic groups. An unconformity at 76 m below surface separates the Tauranga Group from the Te Kuiti Group sediments.*

3. Sample Material And Study Area

Sample	Depth [m]	Stratigraphy	Subcore	Lithology	TOC [%]	R_0 [%]
G004541	18.86-19.02	Taurangura Group	6.3	lignite	45.11	0.29
G002813	19.28-19.43	Taurangura Group		lignite	40	
G004530	25.90-25.97	Taurangura Group		sandstone	0.15	
G004531	25.97-26.03	Taurangura Group		silty sandstone	0.12	
G004532	26.03-26.06	Taurangura Group	9.2	silt	0.28	0.29
G004533	26.06-26.10	Taurangura Group		silty claystone	5.23	
G004534	26.10-26.15	Taurangura Group		lignite/silty clay	4.86	
G004535	26.15-26.17	Taurangura Group		lignite	18.5	
G003579	31.21-31.26	Taurangura Group	11.2	lignite	37.16	0.29
G003580	31.70-31.84	Taurangura Group		lignite	32.81	
G004544	63.41-63.51	Taurangura Group	17.4	lignite	37.87	0.29
G003577	65.27-65.52	Taurangura Group	18.2	lignite	41.79	0.29
G003578	65.52-65.56	Taurangura Group		lignite	40.48	
		unconformity				
G003575	130.38-130.68	Te Kuiti Group	29.5	coal	58.29	0.39
G003574	130.68-130.98	Te Kuiti Group		coal	57.72	
G003576	142.52-142.80	Te Kuiti Group	32.2	coal	59.58	0.39

Table 3.3.: *Selected samples from DEBITS-1 core, including depth [m below surface], stratigraphic groups, subcore number, lithology, TOC content and vitrinite reflectance data. TOC content and R_0 data were provided by R. Sykes (GNS Science, NZ).*

sample set of organic matter rich layers throughout the whole DEBITS-1 well to investigate the potential low molecular weight organic acids (LMWOA) distribution from OM rich layers as well as to examine the LMWOA availability close to OM rich sedimentary layers.

To investigate the potential of different lithologies to provide substrates for deep microbial life, 6 samples from the DEBITS subcore 9.2 representing a transition from organic carbon poor (sandstone) to organic carbon rich (lignite) lithologies were selected. These samples are part of the Puketoka Formation (Late Pliocene to Early Pleistocene). Furthermore, five Tauranga

Group lignite samples were taken from subcores 11.2, 17.4 and 18.2 belonging to the Whangamarino Formation (Latest Miocene to Late Pliocene).

Additionally, two Tauranga Group samples (G004541 and G002813, subcore 6.3, Puketoka Formation) were chosen to contribute to the investigation of the maturity related feedstock potential of the NZ coals reported in chapter 5 representing low mature lignite samples.

The Te Kuiti Group samples belong to the Renown Seam (G003575, G003574, subcore 29.5) and the Kupakupa Seam (G003576, subcore 32.2), both being part of the Waikato Coal Measures Formation.

3.2. Geology of the sedimentary basins of New Zealand

The rocks forming the pre-Cenozoic basement of New Zealand can be grouped into three tectonostratigraphic superterranes. They are known as the Western Province, the Median Tectonic Zone and the Eastern Province (Figure 3.1) (Bradshaw, 1989; Muir et al., 2000). The continental growth along the Pacific margin of Gondwana from Cambrian to Early Cretaceous and processes of the continental dispersal during Late Cretaceous can be recognised from the geological record (Muir et al., 2000).

The Western Province is a continental fragment of Gondwana. It's Paleozoic metasedimentary rocks were intruded by granitoids during Devonian, Carboniferous and Cretaceous. Onshore, the Western Province is limited to the NW and the SW parts of the South Island and extends offshore to the NW. A correlation of the Western Province can be found in the Lachlan Fold Belt in SE Australia and in the Northern Victoria Land and Marie Byrd Land in West Antarctica (Cooper and Tulloch, 1992; Muir et al., 1996).

The Eastern Province comprises volcanic rocks, sedimentary sequences and accretionary complexes of Permian and Mesozoic Age, representing con-

3. Sample Material And Study Area

vergent margin tectonics. Onshore the Eastern Province forms the main part of the basement of the North Island and the eastern half of the South Island. The western parts of the Eastern Province (Brook Street, Murihiku, Matai) are thought to have correlatives in NE Australia and New Caledonia. In contrast to the Western Province, rocks of the Eastern Province have not been identified in West Antarctica (Bradshaw et al., 1997).

The Median Tectonic Zone separates the Western and the Eastern Province. It is a belt of Late Paleozoic and Mesozoic calc-alkaline plutonic rocks with little volcanic and metasedimentary units. These rocks are mainly of Mid Jurassic to Early Cretaceous Age, some smaller parts are of Carboniferous and Triassic Age. The Median Tectonic Zone is thought to be a rest of a magmatic arc system that once developed along the Pacific margin of Gondwana. It forms a belt in the NW Nelson and eastern Fjordland regions of the South Island and extends offshore from Fjordland across the northern half of Steward Island (Figure 3.1) (Allibone and Tulloch, 1997). The Median Tectonic Zone is not found on the North Island but lithologies similar to this (calc-alkaline plutonic rocks) are found in the Taranaki Basin (Mortimer et al., 1997).

Cretaceous and younger sediments that fill the sedimentary basins are important for exploration of hydrocarbons and for the formation of the NZ coal beds. These sedimentary basins are described in the following sections. The geological history of these basins is also described in (Vu, 2008).

3.2.1. Northland Basin

The major important tectonic events for the formation of the Northland Basin can be summarised in three stages which are (1) rifting and subsidence at a passive margin in Middle Oligocene to Early Miocene, (2) convergence, obduction and subduction on a margin in Late Oligocene to Early Miocene and (3) a period of less activity in a back arc and intra-plate setting from Middle Miocene to Holocene (Isaac et al., 1994).

3.2. Geology of the sedimentary basins of New Zealand

The geological setting of the Northland Basin includes the predominantly metamorphosed sandstones and mudstones of the Waipapa Group forming the basement. Ancient unmetamorphosed sediments being found offshore build the Triassic and Early Cretaceous sediments of Murihiku Tarrane. Above, a formation was deposited that comprises Middle to Late Cretaceous coal measures which are equivalent to the coal layers found in the Taniwha and Rakopi Formation of the Pakawau Group in the Taranaki Basin. The western margin of the basement block is overlain by Early Tertiary sediments with some coal measures at the base followed by marine sandstones (Ruatangata Formation) and limestones (Whangarei Formation).

Onshore, the basement is overlain by paralic and shelf facies of Eocene to Oligocene Age forming the Te Kuiti Group. This group comprises the Kamo Coal Measures, glauconites, sandstones and limestones and is overlain by the Northland Allochton, a thick sedimentary and volcanic sequence that is followed by the marine Waitemata Group (Miocene) and Pliocene to recent volcanics (King et al., 1999; Barry et al., 1994).

3.2.2. Taranaki Basin

Tectonics and the complex sedimentary basin evolution are described e.g. in Palmer and Andrews (1993), Killops et al. (1994) and Muir et al. (2000). The tectonic history of the Taranaki Basin can be grouped into two main tectonic phases. There are a passive margin phase from Cretaceous to Oligocene and a convergent margin phase from Late Oligocene to Present. A series of half-grabens is formed from rifting and extension during the passive margin phase and NNE trending sub basins were formed by extension in which a thick section of up to 3 km of Late Cretaceous sediments accumulated. The terrestrial deposits of the Pakawau Group contain the coal layers of the Rakopi Formation. The overlying North Cape Formation formed during Late Cretaceous by the deposition of mudstones and sandstones.

During a period from Late Paleocene to Oligocene, the Taranaki Basin

3. Sample Material And Study Area

was influenced by subsidence and was mainly below sea level. Sandstones were deposited in the southern and the central part of the basin during Paleocene and Eocene and formed the Farewell, Kaimiro and Mangahewa Formations belonging to the Kapuni and the Moa Group. In the northern part of the basin, fine grained and lean shelf sediments formed the Turi Formation.

With the expansion of the modern Australia-Pacific Plate boundary in the Oligocene, the convergent margin phase was initiated. In the Taranaki Basin the subduction began with the basin wide subsidence during Late Oligocene and Early Miocene. In the eastern and southern areas of the basin, the deposition of calcareous siltstones and mudstones formed the Otaraoa Formation, while in the western part of the basin, bioclastic limestones and calcareous mudstones accumulated to form the Tikorangi Formation, both belonging to the Ngatoro Group. In Early Miocene, the hinterland was uplifted and overthrusted along the eastern margin. In the Miocene, the Manganui Formation (mudstones) and the Moko Formation (sandstone) deposited whereas in the northern part of the basin volcanic deposits deriving from the subduction of the Pacific Plate formed the Mohakatino Formation.

During the Late Miocene the compression of the southern Taranaki Basin initiated the inversion of the sedimentary structures and a further uplift of the hinterland lead to the accumulation of sandstones and mudstones of the Matemateaonga Formation in the east and the Giant Foresets Formation in the west.

3.2.3. Waikato Basin

Tectonic events and sedimentation history of the Waikato Basin are reported e.g. by Edbrooke et al. (1994) and Barry et al. (1994). The sediments of the Waikato Basin can be summarised into four lithostratigraphic groups which

3.2. Geology of the sedimentary basins of New Zealand

are the Mesozoic basement, the Paleogene Te Kuiti Group the Early Neogene Waitemata, Mahoenui and Mohakatino Groups and the Late Neogene Tauranga Group.

The Western Province basement rocks are of Early Cretaceous Age and older and belong to two terranes, the Murihiku Terrane in the west and the Waipapa Terrane in the east. The Waipapa sediments of Jurassic Age are mainly made of volcanic derived terrigenous clastics, the Murihiku sediments are of Triassic and Jurassic Age and include conglomerate, lithic sandstone, tuff and siltstone.

The Te Kuiti Group sediments were deposited in the South Auckland area between Late Eocene and Early Miocene. The carbonate content increases upwards and limestone is found near the top in the southern region. The Waikato Coal measures were deposited on top of the Mesozoic Newcastle Group sediments (also see Figure 3.2). The coal measures accumulated in a valley flanked by low basement rocks. The Rotowaro and Huntley coalfields are of Middle to Late Eocene Age, but the coal measures become younger to the north and the south. The sedimentation of the Te Kuiti Group ended during the Earliest Miocene by increased tectonic activity that led to an uplift of the region. This uplift was followed by a time of erosion before a new rapid subsidence lead to the deposition of the Tauranga Group sediments above an unconformity.

The Tauranga Group comprises all post Miocene sediments deposited in the Bay of Plenty, Hauraki Lowland, Hamilton Lowland, Lower Waikato and Manukau Harbour areas. It consists of mainly terrestrial sediments with pumiceous and rhyolitic sands, clay and gravels with imbedded lignite layers. The Tauranga Group is widespread within the Waikato Coal Region.

3.2.4. West Coast Basin

The west coast area of the South Island of NZ was part of Gondwana prior to Cretaceous Age. At the end of Early Cretaceous with the burst of Gondwana,

3. Sample Material And Study Area

the rift basin was formed and terrestrial clastic sediments were deposited. During Late Cretaceous, subsidence of the South Westland area started and continued during the Paleogene. In Middle Eocene, a new plate boundary begun to form and the South Westland drifted away from the Camplell Plateau. The continuous subsidence resulted in the accumulation of marine sandstones, mudstones and limestones that replaced the initial sedimentation of terrestrial sands and coals. At the end of the Eocene and throughout the Oligocene, the entire Western Platform was beneath the sea.

During Late Oligocene and Early Miocene, the Australia-Pacific Plate boundary formed the Alpine Fault (Figure 3.1). Movement along the Alpine Fault shifted the West Coast region northward. The Miocene to Early Quaternary history of the West Coast region is, therefore, influenced by tectonic deformation. This results in an inversion of the north trending grabens and a rapid uplift that caused an increased sedimentation in the adjacent basins.

The old sediments of the Porarari Group are mainly terrestrial clastics deposited during Early Cretaceous. They appear in isolated areas that form the Paparoa Coal Measures in the Westland Basin. Sandstones and coals alternate with lacustrine mudstones and some minor amounts of volcanic tuff. In the Paleogene and Early Eocene, the sedimentation rate was slow and quartz-rich sediments as well as the Brunner Coal Measures formed (King et al., 1999).

During the Late Eocene, the accumulation of dark brown carbonaceous marine mudstones formed the thick Kaiat Formation which is overlain by calcerous muds and limestones. Uplift in Early Miocene resulted in a change of sequences from bathyal siltsones to shallow-water sandstones and terrestrial sequences from the Late Pliocene.

3.2.5. Eastern Southland Basin

In the southwest area, the oldest Tertiary layers are the Mako Coal Measures of Late Eocene to Early Oligocene Age, overlain by predominantly

3.2. Geology of the sedimentary basins of New Zealand

marine limestones and calcerous mudstones of the Winton Hill Formation, the Chatton Formation and the Forest Hill Formation deriving from Early Oligocene to Early Miocene Age (Isaac and Lindqvist, 1990). In the most parts of the Northern and Eastern Southland, the pre-Cenozoic basement is unconformably overlain by the Gore Lignite Measures deriving from Late Oligocene to Middle Miocene. The Gore Lignite Measures contain a sand dominated layer with only minor lignite content overlain by a lignite bearing layer including multiple, laterally persistent seams and an upper layer of mainly sandstones and sandy limestones with only little lignite content. At the Mataura River region in the south of the basin, the lower sand dominated layer is absent but the lignite rich layer is thick.

4. Methods

A detailed view into the structure of the macromolecular matrix and the distribution of specific kerogen-bound compounds is highly relevant for the current study. The main methods used in this study are chemical degradation procedures. Regarding the fact, that low molecular weight organic acids (such as formic or acetic acid) provide important potential substrates, which are suggested to be linked via ester bonds to the kerogen matrix, the main focus in this study is placed on the ester cleavage reaction applied to lignite and coal samples of different maturity. This method provides detailed information about the distribution of low and high molecular weight organic acids and alcohols linked to the kerogen as well as the stability of their linkage with ongoing maturation and structural alteration of the macromolecular matrix. Additionally, ether cleavage reaction was applied to lignites and coal samples to get a further insight into other oxygen bound compounds within the structure of the macromolecular network. For comparison also, open system pyrolysis was applied to the lignites and coals. The flow chart in Figure 4.1 gives an overview about the total sample processing applied in this thesis. Detailed information about the individual steps are presented below.

4. Methods

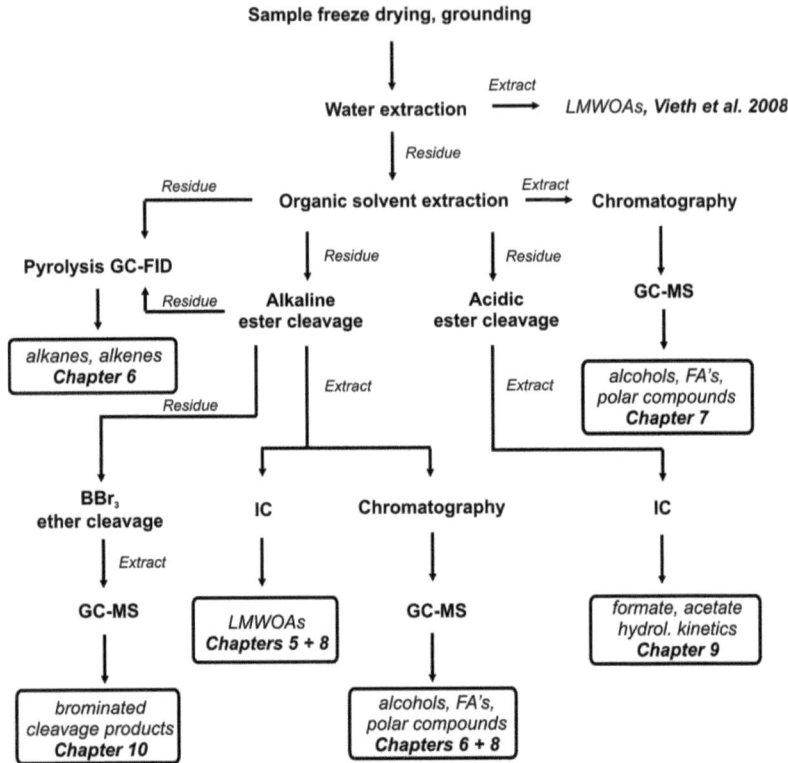

Figure 4.1.: *Flow chart showing the total sample processing applied to the lignite and coal samples in this thesis.*

4.1. Sample preparation

4.1.1. Synthesis of fatty acid ethyl esters (FAEE)

For the evaluation of the developed ester cleavage approach, fatty acids ethyl esters were used. Ethyl acetate and octadecanoic acid ethyl ester were commercially available (Sigma-Aldrich, Taufkirchen, Germany), hexacosanoic and triacontanoic acid ethyl esters were synthesised from the corresponding commercially available free fatty acids.

4.1. Sample preparation

The synthesis of commercially unavailable fatty acid ethyl esters of hexacosanoic acid and triacontanoic acid was performed with some modifications according to the standard procedure published by Blau and Darbre (1993). 10 mg of the fatty acid (purchased from Sigma-Aldrich Chemie GmbH, Buchs, Switzerland) were suspended in 3 ml of dry ethanol in a 10 ml glass vial equipped with a teflon cap. To this suspension, 50 µl of thionyl chloride (Sigma-Aldrich) was added, the vial was closed and heated with warm water until the solution became clear. After cooling to room temperature the solvent, the SO_2 and the HCl were removed under a nitrogen gas stream. The completeness of the reaction and the purity of the products were controlled by GC-MS analysis after methylation. Only ethyl esters were detected indicating complete conversion during the reaction.

4.1.2. Core subsampling, freeze drying and sample grinding

The DEBITS core samples were stored under nitrogen at 4°C. A piece of approximately 5 cm was cut out from a coaly core section. To obtain an uncontaminated sample, an outer section of about 2 cm was cut off and discarded while the inner part of the core section was crushed. These fresh and wet coal samples taken from the DEBITS core were freeze dried for 48 hours using a lyophilisation instrument (Piatkowski-Forschungsgeräte, Munich, Germany). The freeze dried samples were ground to a fine powder with a disc mill for 30 seconds using a steel grinding set. For cleaning, the grinding set was washed with water and rinsed with acetone. Subsequently, an adequate amount of high-purity silica was ground between every new coal sample.

4.1.3. Solvent extraction

In order to investigate the kerogen-bound compounds, compounds from the bitumen fraction had to be removed from the sediment samples by extraction

4. Methods

using organic solvents. Furthermore, the obtained organic solvent extract was used for the investigation of the free compounds (cf. Chapter 7).

For the extraction of the bitumen an azeotropic solvent mixture of acetone (38%), chloroform (32%) and methanol (30%) (Sigma-Aldrich Chemie GmbH, Buchs, Switzerland, GC quality) was used. The extraction was performed in a Soxhlet apparatus for 3 d. An additional Soxhlet extraction using pure methanol (being the solvent of the subsequent hydrolysis) was added for one day. Samples used for the alkaline ester cleavage procedure were extracted for 24h with water prior to the organic solvent extraction in order to remove water extractable free low molecular weight organic acids.

4.1.4. Ester cleavage procedures

For the hydrolysis of kerogen-bound acids, alkaline ester cleavage (saponification) was used (cf. Chapters 5,6 and 8). This reaction was performed using either methanol or water as solvent. Additionally, acid catalyzed ester hydrolysis was used for investigation of the kinetics of a potential pathway for the liberation of kerogen-bound esters in nature (cf. Chapter 9).

Procedure for alkaline hydrolysis in methanol

In a 50 ml round flask, 2 g of the extracted and dried coal sample were suspended in 20 ml of a 2 N KOH solution in methanol and refluxed for 5 h. Afterwards, the mixture was allowed to cool down to room temperature. In the next step, 50 μg of erucic acid and 50 μg of androstanone were added as internal standards to the reaction mixture. The residue was filtered and washed three times with 20 ml of methanol. Methanol was removed under reduced pressure from the alkaline solution. The precipitating residue obtained after evaporation of the solvent was dissolved in 25 ml of water (purified by deionisation and by Millipore™ filtration) and separated into two parts.

4.1. Sample preparation

The first part was acidified using 5 N HCl up to pH 3 and extracted 3 times with 20 ml dichloromethane. The extract was dried over Na_2SO_4 and the solvent was removed under reduced pressure. The remaining products were dissolved in dichloromethane and analysed by GC-MS for the detection of high molecular weight fatty acids (C_12+), after fractionation by column chromatography (see below).

The second part was brought to pH 10 with 5 N HCl, transfered into a volumetric flask and filled up to 20 ml with water. This sample was used for IC analysis to detect low molecular weight organic acids.

The residue was again suspended in 20 ml methanol and acidified with diluted HCl up to pH 3. The mixture was than filtered and the residue was washed with methanol to remove the acid completely and dried before further use (see ether cleavage procedure, Chapter 4.1.5).

Procedure for alkaline hydrolysis in water

Aproximately 2 g of the extracted and dried coal sample were placed into a 10 ml vial and suspended in 5 ml of 2 N aqueous KOH. The samples were placed in an oven and heated at 120°C overnight. After cooling to room temperature, the residue and the alkaline aqueous extract were separated using centrifugation. The extract was than split into two parts. The first part was acidified using 5 N HCl up to pH 3 and extracted 3 times with 20 ml dichloromethane. The combined organic solvent extracts were dried over Na_2SO_4 and the solvent was removed under reduced pressure. The remaining products were used for GC-MS analysis of the high molecular weight fatty acids. The second part was brought to pH 10 with 5 N HCl, transfered into a volumetric flask and filled with Millipore™ filtered water up to 20 ml. This sample was used for IC analysis of the low molecular weight organic acids.

4. Methods

Procedure for acidic hydrolysis in water

Approximately 300 mg of the pre-extracted (*cf.* Section 4.1.3) and dried coal samples were weighed into a 20 ml head space vial and mixed with 5 ml of diluted HCl (pH 3). The vials were closed and placed into an oven at different temperatures for different time periods (see Chapter 9 for details). The heated samples were removed from the oven, cooled to room temperature in a water bath and filtered using nitrogen overpressure. The aqueous extract was transfered into a volumetric flask and filled up to 10 ml. These samples were used for IC analysis of the low molecular weight organic acids.

Fractionation of saponification products by liquid column chromatography

The fractionation of the high molecular weight ester cleavage products was performed by column chromatography. Alkaline treated silica gel was used as solid phase. 3 g of silica gel (Merck, K 60) were suspended in 20 ml of 1 N KOH (in MeOH) and isopropanol (1:1, v/v) and stirred for 10 min. This so-treated silica gel was then transfered into a glass column and conditioned with 300 ml of dichloromethane. The chromatography was performed using nitrogen overpressure of approximately 3 bars. While the fatty acids are precipitating as their potassium salts on the silica column, the neutral fraction was eluted with 250 ml dichloromethane. Afterwards, the acid fraction was liberated and eluted with 250 ml 2% formic acid in dichloromethane. The solvent was removed and the remaining fractions were used for GC-MS analysis. The acid fraction was derivatised with diazomethane to form methylesters, the neutral fraction, containing the alcohols, was treated with N-Methyl-N-(timethylsilyl)fluoroacetamide (MSTFA) for derivatisation of the hydroxy groups into TMS derivatives.

4.1.5. Ether cleavage procedure

For the investigation of ether-bound compounds, a reaction involving ether cleavage was applied (*cf.* Chapter 10) following the previously described ester cleavage procedure.

Approximately 150 mg of the dried residue after ester cleavage were transfered into a 10 ml vial. 5 ml of BBr$_3$ solution (1 M in DCM, Alfa Aesar GmbH & Co.KG, Karlsruhe, Germany) were added and the vial was closed. The suspension was treated in an ultrasonic bath for 2 h, then stirred for 24 h at room temperature and ultrasonicated for an additional 2 h period. After cooling to room temperature, 2 ml of diethyl ether were carefully added to the stirred mixture to destroy the excess of reactant. The suspension was filtered using Whatmann glas fiber filters (poresize 0.7 μm) and the residue was washed three times with 5 ml of diethyl ether. The combined organic washings were dried with Na$_2$SO$_4$ and the solvent was reduced to a volume of 2 ml at ambient pressure using a TurboVap® device.

4.2. Instrumental analysis

4.2.1. Open system pyrolysis (pyrolysis-GC-FID)

For open system pyrolysis-gas chromatography (pyrolysis-GC) analysis, approximatly 4 mg of coal were placed into the central part of a glas tube of 26 mm length and 3 mm in diameter. The remaining volume was filled with quartz wool cleaned at 630°C for 30 min. The pyrolysis was realised using a *Quantum MSSV-2 Thermal Analyzer* (for details of the method, see Horsfield, 1989) interfaced with an *Agilent GC-6890A* gas chromatograph.

The sample was heated in a flow of helium to 300°C to remove the volatile material. Subsequently, the samples were pyrolysed using a temperature program from 300°C up to 600°C with a rate of 50°C/min and held for 3 min isothermal. Pyrolysis products were cryogenically trapped with a liquid nitrogen trap and afterwards released by fast heating up to 320°C (held for 10 min). Product analysis was done with a dimethylpolysiloxane capillary column (HP-Ultra 1; 50 m length, 0.32 mm diameter, 0.52 μm film thickness) connected to a flame ionisation detector (FID). Helium was used as carrier gas. The temperature of the gas chromatograph was programmed from 30°C to 320°C at a rate of 5°C/min, followed by an isothermal phase of 35 min.

4.2.2. Gas chromatography-mass spectrometry (GC-MS)

GC-MS analysis was performed using a *Trace GC Ultra* linked to a *DSQ* mass spectrometer both from Thermo Electron Corporation (Thermo Fisher Scientific Inc.,Waltham, MA, USA). The GC-oven temperature was set to 50°C (1 min isotherm), heated with a heating rate of 3°C/min to 310°C, and kept isothermal for 30 min at 310°C. The GC was equipped with a *PTV Splitless-injector* heating from 50°C to 300°C with a rate of 10°C/s. For compound separation a fused silica gel *SGE BPX 5* with 50 m length, a diameter of 0.22 mm and 0.25 μm film thickness was used. Helium was used as carrier gas. The MS was operated in the electron impact (EI) ionisation mode at 70 eV and the source temperature was set to 230°C. Full scan spectra were recorded from 50 to 650 amu with 2.5 scans/sec. The amounts of aliphatic fatty acids were calculated from the m/z 74 mass trace according to an internal standard, the amounts of the aliphatic alcohols were calculated from the m/z 75 mass chromatogram.

4.2.3. Ion chromatography (IC)

IC analysis was performed on a *Dionex ICS 3000* instrument (Dionex Corporation, Sunnyvale, CA, USA) with a KOH eluent generator equipped with an *ASRS Ultra II 2 mm* suppressor and Dionex conductivity detector. The column used was *AS11HC* (2 x 250 mm) with an *AG11HC* pre-column. Column temperature was set to 35°C. Quantification of the components was done using an external standard calibration curve. Samples were eluted using a KOH solution of varying concentrations over time. The initial KOH concentration was 0.5 mM, held for 8 min. After 10 min, a concentration of 15 mM KOH was reached and kept constant for 10 min. After 30 min analysis time, a concentration of 60 mM KOH was reached, followed by a rapid increase to 100 mM reached after 30.2 min analysis time. After 32 min, the KOH concentration was again at the initial level of 0.5 mM and kept for an additional 15 min to equilibrate the system. For quantification of organic acids, standards containing all investigated compounds were measured in different concentrations every day.

4.2.4. Laboratory equipment

Samples and chemicals were weighed using a *Sartorius Genius ME215S* (Sartorius AG, Göttingen, Germany). Centrifugation was performed using a *Sigma 6415* centrifuge (Sigma, Osterode am Harz, Germany). A *Simplicity 185* apparatus (Millipore Corporation, Molsheim, France) was used to remove the organic compounds from the deionised water. Removal of organic solvents from the sample extracts or fractions was performed either under reduced pressure using a *Büchi Rotovapor R-205* (Büchi Labortechnik GmbH, Essen, Germany) with a water bath of 40°C or using a *TurboVap 500* (Caliper Life Sciences Inc., Hopkinton (MA), USA) at ambient pressure.

4.3. Software

GC-MS analysis and GC-MS data evaluation was performed using the software *Xcalibur* (Thermo Electron Cooperation). The pyrolysis-GC-FID data were evaluated using the *Chem Station* software (Agilent Technologies), for

4. Methods

IC analysis the program *Chromestar* was used. Calculations were done with *Microsoft Excel*, chemical structures were drawn using *CS ChemDraw Ultra 6.0*. Graphs and figures were designed with *Microsoft Excel*, *Grapher 6* and *CorelDRAW 12*. This thesis was edited with $\LaTeX$ using *MiKTeX 2.7* and the *TeXnicCenter* editor.

5. Ester Bound Low Molecular Weight Organic Acids Linked To The Kerogen

This chapter was published in:

Glombitza, C.; Mangelsdorf, K.; Horsfield, B. (2009). A novel procedure to detect low molecular weight compounds released by alkaline ester cleavage from low mature coals to assess its feedstock potential for deep microbial life. Organic Geochemistry, 40(2):175-183.

5.1. Introduction

It was long thought that life on Earth was restricted to the surface and the upper few meters of sediment (Morita and ZoBell, 1955). In recent years the finding of deep microbial life in the subsurface opens the view to a fascinating new topic in today's microbiology and geoscience research (Kerr, 1997; Whiteman et al., 1998; Parkes et al., 2000; Horsfield et al., 2007). With the discovery of an ubiquitous deep biosphere on Earth, inevitably the question on its potential carbon and energy sources arises.

In environments with little buried organic matter, for instance in igneous rocks like basalts and granites, lithoautotrophic microbial communities are able to synthesise small organic compounds from inorganic sources

5. Ester Bound Low Molecular Weight Organic Acids Linked To The Kerogen

like radiolytically generated hydrogen gas and inorganic carbon dioxide, and these simple organic compounds can be utilised by other microorganisms. Therefore, such microbial communities form deep subsurface microbial ecosystems being independent from photosyntetically produced substrates (Stevens and McKinley, 1995; Fredrickson and Onstott, 1996; Lin et al., 2005, 2006; Sherwood-Lollar et al., 2006). In marine and terrestrial sedimentary basins, the most obvious energy source is, of course, buried organic matter (Froelich et al., 1979; Stetter et al., 1993; D'Hondt et al., 2004). Caused by biotic but also abiotic alteration, the sedimentary organic matter undergoes a permanent restructuring process with increasing diagenesis, thereby releasing nitrogen, sulphur, and oxygen containing compounds (Sinninghe Damsté et al., 1989a; Eglinton et al., 1992; Béhar et al., 1995; Putschew et al., 1998; Vandenbroucke and Largeau, 2007). Thus, organic carbon rich lithologies could act as potential "feeder" lithologies for the deep biosphere (L'Haridon et al., 1995; Krumholz et al., 1997; Horsfield et al., 2006).

Elucidation of the structural composition of sedimentary organic matter was and still is one of the most intriguing and challenging topics in organic geochemistry. To improve the understanding of the molecular composition of kerogen, various analytical techniques were used such as nuclear magnetic resonance spectroscopy (NMR) (Derbyshire et al., 1989; Burgess and Schobert, 1998; Suggate and Dickinson, 2004; Werner-Zwanziger et al., 2005), infrared spectroscopy (IR) (Robin, 1975; Rullkötter and Michaelis, 1990; Lis et al., 2005), pyrolysis gas chromatography-mass spectrometry (Pyrolysis GC-MS) (Larter and Douglas, 1982; Larter and Horsfield, 1993; Khaddor et al., 2002; Killops et al., 2002; Sykes and Snowdon, 2002) and selective chemical degradation reactions (Schouten et al., 1993; Schaeffer et al., 1995; Höld et al., 1998; Putschew et al., 1998; Schaeffer-Reiss et al., 1998). These powerful analytic tools provided various information about the compositional structure of kerogen and its maturation history (Vandenbroucke and Largeau, 2007). Especially, for the elucidation of the different

5.1. Introduction

bonding types, selective chemical degradation reactions have been demonstrated to be a highly sophisticated method (Rullkötter and Michaelis, 1990). Main decomposition steps are the cleavage of ester, ether, sulphur (Hofmann et al., 1992; Schaeffer et al., 1995; Höld et al., 1998) and carbon-carbon (Stock and Tse, 1983; Stock and Wang, 1986) bonds, providing valuable information on the internal bonding structures within the kerogen. Additionally, these techniques allowed insights into the chemical and structural variation within the kerogen during diagenetic alteration with increasing maturity.

Important substrates for feeding microbial life are low molecular weight organic acids (Sansone and Martens, 1981, 1982; Sørensen et al., 1981; Jørgensen, 1982). Therefore, the cleavage of ester bound compounds is a promising tool to examine the potential of sedimentary organic matter of different levels of maturity to release organic acids which can potentially act as a feedstock for deep microbial life. The commonly used and well established procedure for the ester cleavage is alkaline hydrolysis (Putschew et al., 1998; Schaeffer-Reiss et al., 1998). However, due to the experimental preparation procedure, the conventional ester cleavage methods are mainly focussed on the long chain, non-volatile ester cleavage products.

In this paper, a novel analytical approach is presented to detect low molecular weight (LMW) organic acids from alkaline ester cleavage reaction of coal samples of different maturity. Simultaneously, analysis of the high molecular weight (HMW) product compounds (alcohols and fatty acids) can be performed using this approach. However, results of this investigation will be published elsewhere. Using the developed approach, a series of New Zealand (NZ) coal samples of continuous maturity from lignite to high volatile bituminous coal rank was investigated to determine the potential of these coals to release LMW organic acids (such as formic acid, acetic acid and oxalic acid) from macromolecular coal matrix forming a potential feedstock for deep microbial life.

5.2. Sample material

The North and South Island of New Zealand are home to coals covering a broad maturity range of almost continuous maturity (R_0: 0.25 to 3.0%) from Cretaceous to Tertiary age in freshly exposed coal facies, coal mines and drill cores (Figure 3.1) forming the so called New Zealand Coal Band (Suggate, 2000; Killops et al., 2002; Sykes and Snowdon, 2002). Therefore, the New Zealand Coal Band provides one of the best maturity rank series in the world to examine natural maturation processes in coals. For the current study being part of the international DEBITS (Deep Biosphere in Terrestrial Systems) project, we selected 12 samples from different coal locations on the North and South Island. Two of these samples were taken from the DEBITS-1 well, which was drilled within the scope of the DEBITS project on a meadow near the small village of Huntly on the North Island of New Zealand. The coals selected for this study cover a maturity rank range from lignites to high volatile bituminous coals (R_0: 0.28 to 0.80%) and, therefore, represent different feeding potentials. The sample material was part of several coalfields from four different coal bearing basins (Figure 3.1, Table 5.1): one sample from the Eastern Southland Basin being in the lignite rank range (R_0: 0.28%), two coal samples from the DEBITS-1 well in the Waikato coal area also in the lignite rank range (R_0: 0.29%), three additional samples from the Waikato Basin being in the sub-bituminous rank range (R_0: 0.40 to 0.44%) and four coals from the West Coast Basin as well as two from the Taranaki Basin representing all high volatile bituminous coals (R_0: 0.52 to 0.80%). The selected samples cover a time period from Late Cretaceous to Pleistocene age (Table 5.1).

5.2. Sample material

Sample	Basin	Coalfield / Sample location	Formation	Age	R_o %	TOC %	Formate mg/gSed	Formate mg/gTOC	Acetate mg/gSed	Acetate mg/gTOC	Oxalate mg/gSed	Oxalate mg/gTOC
G001978	Eastern Southland	Waimumu	Gore Lignite Measures	Oligoncene - Miocene	0.28	52.44	16.5	31.4	3.4	6.5	0	0
G004541		DEBITS-1 (18.94 m)	Puketoka	Pliocene - Pleistocene	0.29	45.11	7.9	17.5	1.8	3.9	1.8	3.4
G004544	Waikato Basin	DEBITS-1 (54.46 m)	Whangamarino	Miocene - Pliocene	0.29	37.87	10.8	28.7	1.2	3.3	2.2	5.7
G001982		Rotowaro	Waikato Coal Measures	Eocene - Oligocene	0.40	54.81	12.9	23.5	3.2	5.9	0	0
G001983		Maramarua			0.41	53.82	6.1	11.4	1.6	3	1.6	3
G001980		Rotowaro			0.44	61.16	4.5	7.4	1.4	2.2	1.5	2.5
G001996		Reefton			0.52	67.85	4.7	7	1.1	1.6	1	1.5
G001995	West Coast Basin	Reefton	Brunner Coal Measures	Eocene	0.52	67.76	4.3	6.4	1	1.5	0.8	1.2
G001989		Buller			0.69	73.33	2.5	3.5	0.7	0.9	0.7	1
G001993		Garvy Creek			0.76	77.73	1.6	2.1	0.5	0.7	0.6	0.7
G001994	Taranaki Basin	Paturau River	Rakopi	Late Cetaceous	0.61	63.99	1.4	1.8	0.2	0.3	0.5	0.8
G001991		Ohanga-2 well	Mangahewa	Eocene	0.80	73.83	3.2	4.4	0.8	1.1	0.6	0.8

Table 5.1.: Selected samples for the investigation of kerogen bound formate, acetate and oxalate including information about basin, coalfield, formation, age, vitrinite reflectance and TOC (data provided by R. Sykes, GNS Science (NZ)) and the detected amounts of formate, acetate and oxalate.

5.3. Experimental approach

As mentioned above, the commonly used procedures for alkaline ester cleavage are usually restricted to the recovery of high molecular weight cleavage products. In the modified approach, described below, we combine the conventional procedure covering the high molecular weight (HMW) products with a new procedure to detect the low molecular weight (LMW) products using ion chromatography (IC) (Figure 5.1).

5.3.1. Sample preparation

In order to remove the bitumen fraction from the coal samples, the freeze dried and ground coal samples were pre-extracted with water using a Soxhlet extraction for two days, followed by a second Soxhlet extraction for three days with an azeotrope solvent mixture of acetone (38%), methanol (30%) and chloroform (32%). Prior to the ester cleavage reaction, the coal sample was additionally extracted with the reaction solvent (methanol) for one day to ensure that the compounds obtained after alkaline hydrolysis indeed represent the ester bound lipid fraction.

The cleavage reaction was conducted according to the method described by Höld et al. (1998). Two grams of the pre-extracted coal sample were suspended in 20 ml 1 N KOH solution in methanol and refluxed for 5 h. Optionally the cleavage reaction can also be performed in sealed vials placed in an oven at 100°C for 18 h. Afterwards, the suspension was cooled to room temperature and 50 μg androstanone and 50 μg eruic acid were added as internal standards to trace losses in the yield of the target compounds during subsequent sample treatment and to quantify the HMW cleavage products using gas chromatography-mass spectrometry (GC-MS). To remove the coal matrix, the suspension was filtered, whereas the residue was washed three times with 50 ml of methanol. Subsequently, all filtrates were combined and split into two proportions. The first half was used for the investigation

5.3. Experimental approach

of the LMW ester cleavage products, the second half for the HMW products (cf. Chapter 6) using the conventional analytical method described by Höld et al. (1998).

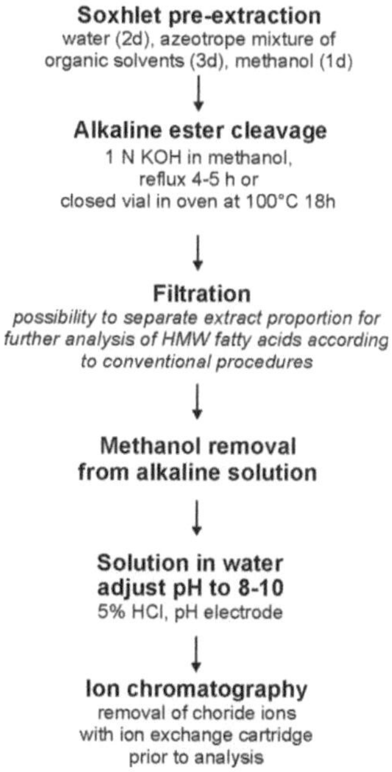

In the conventional method, the next steps would be acidification of the filtrate with aqueous HCl to pH 3 to transform the alkali salts of the carboxylic acids into their corresponding acids, extraction of the acids with dichloromethane, removal of the organic solvent to dryness, methylation of the acids and preparation of a solution for GC-MS measurement. As far as LMW organic acids are concerned, this method is inadequate because small organic acids like formic and acetic acid are highly volatile and relatively polar. Thus, recovery problems for these compounds occur during the extraction of the water-methanol phase with dichloromethane and the subsequent removal of the solvent supporting the evaporation of the small volatile organic acids. Removal of the solvent is necessary, because ions have to be dissolved in water for ion chromatography. Additionally, the LMW organic acids cannot be measured directly from the reaction solution, because methanol and high amounts of hydroxyl ions interfere with the detection of small organic acids in the ion chromatography.

Soxhlet pre-extraction
water (2d), azeotrope mixture of organic solvents (3d), methanol (1d)
↓
Alkaline ester cleavage
1 N KOH in methanol,
reflux 4-5 h or
closed vial in oven at 100°C 18h
↓
Filtration
possibility to separate extract proportion for further analysis of HMW fatty acids according to conventional procedures
↓
Methanol removal from alkaline solution
↓
Solution in water adjust pH to 8-10
5% HCl, pH electrode
↓
Ion chromatography
removal of choride ions with ion exchange cartridge prior to analysis

Figure 5.1.: *Experimental procedure for analysis of LMWOAs released from coals by alkaline ester cleavage reaction.*

To overcome this problem in the proportion of the LMW ester cleavage product analysis, methanol was directly removed from the alkaline cleavage solution under reduced pressure. The organic acids precipitated as their corresponding potassium salts. Subsequently, the potassium salts were dissolved in water and the solution was adjusted to pH 8-10 with an aqueous solution of 5% HCl. Afterwards, the slightly alkaline solution was transferred into a volumetric flask and water was added up to 20 ml. This sample was used for ion chromatographic analysis of the LMW organic acids.

5.3.2. Ion chromatography

Prior to IC analysis, the samples were diluted with water (1:10) and an ion exchange cartridge (Dionex On Guard II Ag/H) was used to remove high chloride contents. The cartridge contains a styrene based sulfonic acid that replaces the cloride ions by sulfate ions. In contrast to the chloride ions the sulfate ions do not interfere in the chromatogram with the target compounds because of their longer retention time. Test experiments showed that the use of the applied ion exchange cartridge had no effect on the organic acid concentration in the samples. All samples were analysed in replicate using a Dionex ICS 3000 ion chromatograph with a KOH eluent generator equipped with an ASRS Ultra II 2 mm suppressor and a Dionex conductivity detector. For anion separation the analytical column AS11HC (2 x 250 mm, Dionex Corp.) with a pre-column AG11HC was used at a temperature of 35°C. Samples were eluted using a KOH solution of varying concentration over time. The initial KOH concentration was 0.5 mM, held for 8 min. After 10 min, a concentration of 15 mM KOH was reached and kept constant for 10 min. After 30 min analysis time, a concentration of 60 mM KOH was reached, followed by a rapid increase to 100 mM reached after 30.2 min analysis time. At 32 min, the KOH concentration was again at the initial level of 0.5 mM and kept there for an additional 15 min to equilibrate the system. For quantification of organic acids, standards containing formic acid, acetic acid (suprapur, Merck, Germany) and oxalic acid (p.a., Merck, Germany) were measured in different concentrations every day. Standard deviation of sample and standard quantification is below 10% (determined from replicate analysis).

5.3.3. Method evaluation

To evaluate the general applicability of the presented methods especially in terms of compound recovery, test cleavage experiments were conducted using propyl formate, ethyl acetate (both Sigma-Aldrich, Taufkirchen, Germany), C_{18}, C_{26} and C_{30} fatty acid ethyl esters in known amounts of 2, 4, 6, 8 and 10 mg. However, these test experiments do not consider any adsorption effects of the coal matrix. The cleavage reactions were performed in methanol under reflux as described above. For comparison, another cleavage reaction method was tested, where the reaction was conducted in a closed glass vial for 18 h in an oven using 5 ml 1N KOH in methanol (oven temperature 100°C) and 5 ml 1N KOH in water (oven temperature 120°C).

While propyl formate, ethyl acetate and octadecanoic acid ethyl ester was commercially available (Sigma-Aldrich, Taufkirchen, Germany), the longer chain fatty acid ethyl ester standards with 26 and 30 carbon atoms were synthesised from hexacosanoic and triacontanoic acid in a 10 ml sample vial according to a procedure published by Blau and Darbre (1993). To optimise the yields obtained by this method the closed reaction vials containing the respective acid, ethanol and thionyl chloride were additionally heated in water bath at 40°C until the solution became clear. The purity of the products was controlled by GC-MS analysis.

5.4. Results and discussion

5.4.1. Evaluation of the analytical method for the recovery of low molecular weight organic acids

The test cleavage experiments with different standard amounts show a retrieval of 69-79% for formate and ca. 73-85% for acetate using methanol as the reaction solvent under reflux conditions for 5 h (Figure 5.2). Conducting the experiments in a closed vial placed in an oven at 100°C for 18 h using methanol as solvent leads to a recovery of 63-84% of formate and 65-95% of acetate. In contrast the same experimental assembly using water as solvent leads to a recovery of 75-82% of formate and 84-88% of acetate. The retrieval of formate and acetate during the test experiments with methanol under reflux and water in a vial are relatively constant over the whole range of different standard amounts indicating that at least in the chosen range different compound amounts have no significant effect on the cleavage reaction. However, the recovery of the test experiments with water in a vial is slightly better. The recovery of formate and acetate in the methanol in a vial experiment using 8 mg and 10 mg is distinctly lower than in the experiments with 2 mg, 4 mg and 6 mg. It can be speculated that in the vial experiment side reactions like decarboxylation and chemical degradation are responsible for the lower recovery, which might be expected to be more significant at higher compound concentrations. However, this was not observed in the water in a vial experiment and also not in the methanol reflux experiment. Thus, these lower yields might be more likely related to losses during sample processing.

Although the cleavage reaction with water shows slightly higher recoveries, we decided to use the method with methanol under reflux for 3 reasons: (1) the mixing of the coal matrix with methanol and, therefore, with the reagents was much better than with water, an effect not considered in our test experiments with only 5 selected standard compounds. (2) The separation of the reaction solution from the coal matrix by filtration is much more time consuming when water is used as the solvent. (3) Long chain fatty acid experiments indicate that the ester cleavage reaction is not complete using water as solvent which is inadequate when the investigation of both low and high molecular weight fatty acids is intended. In a test experiment C_{18}, C_{26} and C_{30} fatty acid ethyl esters were treated with 1 N KOH solution in water under reflux for 5 h. Non-cleaved ethyl esters were detected in DCM extract of the reaction mixture, whereas the amount of ethyl esters increased with

5.4. Results and discussion

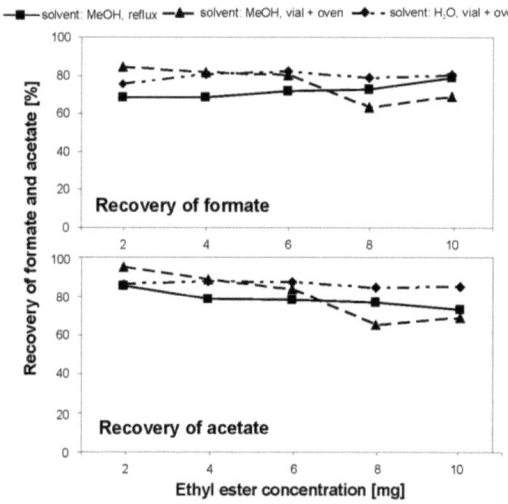

Figure 5.2.: *Recovery of formate and acetate after ester cleavage of its ethylated congeners using different analytical ester cleavage conditions: with methanol (MeOH) under reflux, with MeOH in a closed vial in an oven at 100° C and with water in an closed vial in an oven at 120° C.*

increasing chain length (C_{18}: 6.1%, C_{26}: 9.3% and C_{30}: 11.1%). These findings might reflect differences in the solubility of the investigated fatty acids in water becoming poorer with the increasing aliphatic character of longer chain fatty acids.

A problem with the use of methanol as a solvent could be the transesterfication reaction (Otera, 1993) leading to methyl esters instead of corresponding potassium salts of the acids. This would mean that a proportion of the LMW organic acids could be lost as highly volatile methyl esters during the removal of methanol under reduced pressure. Transesterification takes place when the attacking nucleophile is a methanolate instead of a hydroxyl ion. Methanolate ions simply form when hydroxyl ions are solved in methanol. For instance, Caldin and Long (1954) showed for a solution of 0.1 N NaOH in 99% ethanol that 96% of the solved base was ethanolate.

5. Ester Bound Low Molecular Weight Organic Acids Linked To The Kerogen

Transesterfication reaction is used e.g. for the production of biodiesel from vegetable oils (Schuchardt et al., 1998). Comparing the cleavage experiment curves in figure 3 it can be recognised that the recoveries for the water experiments (where no transesterfication can take place) are slightly better than for the methanol experiments. However, this difference is only small and might also derive from operational variations instead of transesterfication. An indication that transesterfication is not a significant side reaction in the applied procedure is that no ethyl and methyl esters of long chain fatty acids (C_{18}, C_{26} and C_{30}) can be detected in the GC run of the organic solvent extracts directly obtained from the alkaline ester cleavage solution without derivatisation (methylation). The lack of ethyl esters indicates complete saponification and the lack of methyl esters that transesterfication plays not a significant role. After methylation of the extract with diazomethane the saponified fatty acids can be detected as their methyl esters.

To understand why transesterification might not play a role, it has to be noted that every possible transesterification product can again be attacked by another nucleophile. This would result again in a methyl ester in the case of another methanolate or if the nucleophile is a hydroxyl ion into a free acid or under alkaline conditions into the corresponding potassium salt. The deprotonated acid is not reactive anymore for a nucleophilic attack due to the mesomeric stabilisation of the carboxylic ion (Exner, 1988). Thus, the deprotonated acid forms a trap for the competing chemical reactions. In contrast to industrial processes like the production of biodiesel by transesterfication (Schuchardt et al., 1998), where KOH is only used in catalytic amounts (1-2 mol%) compared to the initial esters (vegetable oils), in the applied geochemical reaction scale the nucleophile is used in a large excess and, therefore, might cause complete saponification.

Thus, the observed losses of the target substances in the test experiments (Figure 5.2) might more likely be the result of product losses during the diverse sample treatment steps in the applied complex analytical method. Finaly, it has to be mentioned that the test experiments do not consider any adsorbtion effects of the coal matrix concerning the ester cleavage products which might occur in natural samples.

5.4.2. LMW organic acids linked to NZ coals of different maturity

The 12 NZ coal samples of different maturity (Table 1) were investigated applying the developed analytical approach. Using ion chromatography the ester cleavage products formate, acetate and oxalate were detected in significant amounts in almost all samples investigated (Figure 5.3). The main compound released during the ester cleavage experiment is formate, followed by acetate and oxalate. During the diagenetic phase, the amount of ester linked formate rapidly decreases with increasing maturation of the coals. During early catagenesis the decrease slows down and remains low from a maturity level of about 0.6% R_0, the onset of the oil window of the coals from the New Zealand Coal Band (Sykes and Snowdon, 2002). For acetate and oxalate the same trend is observed, although oxalate was not detected in all samples investigated. The higher variation in the amounts of the released low molecular weight fatty acids during the diagenetic stage is most likely the result of a higher variability in the organo facies of the individual coals of lower maturity, which even might occur in coal samples from the same coalfield. Sykes and Snowdon (2002) also show a high variability in the bitumen index data for the New Zealand coals of lower maturity indicating differences in the organo facies of the investigated coals.

Although there are some variations, an overall trend of a decreasing amount of the ester cleavage products formate, acetate and oxalate can be observed with increasing maturity and, thus, indicates a continuous loss of kerogen linked small organic acids during maturation of the organic matter. The lignite sample from the Waimumu coalfield (Eastern Southland Basin) showed the highest amounts of formate and acetate with up to 31.4 mg/g TOC and 6.4 mg/g TOC, respectively, but no oxalate. The highest amounts

5. Ester Bound Low Molecular Weight Organic Acids Linked To The Kerogen

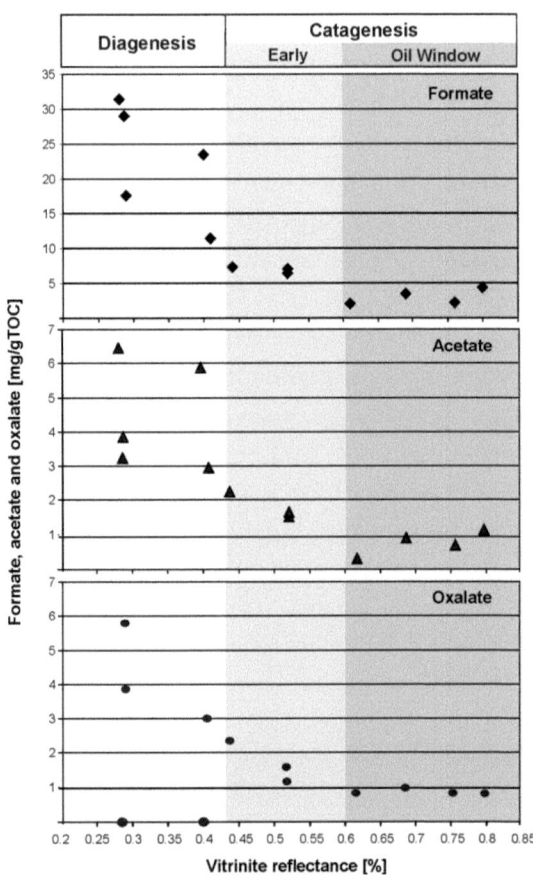

Figure 5.3.: *Formate, acetate and oxalate concentrations liberated by ester cleavage reaction from New Zealand coals of different maturity. The onset of early catagenesis was deduced from Vu et al. (2009). The zone between 0.5 and 0.6% R_0 is suggested to be the onset where temperature conditions might become incompatible with microbial life.*

of oxalate were detected in the sample taken from the DEBITS-1 core from the Wahangamarino formation at 64.46 m depth (Waikato Basin). The subbituminous coals from the Waikato Basin show still high but strongly decreasing amounts of formate and acetate and oxalate (exception sample G001982) with increasing maturity. This trend can also be followed in the high volatile bituminous coals in the West Coast Basin until finally reaching low values of about 2-4 mg/gTOC formate and 0.1-1 mg/gTOC acetate and oxalate at maturity levels higher than 0.6% R_0 for coal samples from the West Coast and Taranaki Basins. Thus, the data indicate that the main decrease of LMW organic acids is associated with diagenetic structural alteration processes characterised by the loss of oxygen containing compounds.

Temperature is thought to be a limiting factor for the occurrence of microbial life in the deep subsurface. Deduced from investigations in petroleum reservoirs, to date temperatures between 80 and 90°C are suggested to form the limit temperatures for deep microbial activities (Connan, 1984; Wilhelms et al., 2001; Head et al., 2003). Using the kinetic model of Sweeney and Burnham (1990) with an average heating rate for geological systems of 2°C per Ma, for the current study, temperatures between 80 and 100°C are calculated for the maturity level between 0.5% and 0.6% R_0. Thus, conditions being incompatible with microbial life are suggested to commence during the later stage of early catagenesis.

The overall decrease of oxygen bearing compounds coincides with observation on diagenetic restructuring processes of kerogen obtained by other analytical methods. For instance, the evolution pathway of buried organic matter plotted in a van Krevelen diagram indicates as well an enhanced loss of oxygen containing compounds during diagenetic maturation by the preferential decrease of the atomic O/C ratio relative to the H/C ratio (Tissot and Welte, 1984b). For the selected sample material of the NZ coals, the loss of oxygen was shown in a van Krevelen diagram by Vu (2008). Furthermore, infrared spectroscopy of kerogen samples with increasing maturity reveals the loss of compounds containing a CO moiety as indicated by the decrease

5. Ester Bound Low Molecular Weight Organic Acids Linked To The Kerogen

of the CO vibration at ca. 1700 cm^{-1} (Robin, 1975; Tissot and Welte, 1984b; Rullkötter and Michaelis, 1990; Lis et al., 2005). Concomitantly, these infrared analysis show that the decrease in the CO vibration continues up to 3% R_0 as reported for the Sarufutsu coals (Takeda and Asakawa, 1988). This might indicate that the release observed in Figure 5.3 will proceed slowly due to thermogenic degradation processes at maturity levels higher than those covered by the current study.

5.4.3. Feedstock potential of the NZ coals of different maturity

The data presented suggest that there is a progressive release of LMW organic acids with ongoing maturation. Thus, LMW organic acids are permanently liberated from the coals into the surrounding sediments. Therefore, the LMW organic acids linked to the coal matrix form a potential feedstock for deep microbial life. To estimate this feedstock potential, CO_2 respiration rates of 10^{-6} to 10^{-3} mmol per litre and year, reported by D'Hondt et al. (2002) for deep-aquifer systems (Table 5.2), were compared to the amounts of LMW organic acids released from the NZ coals of different maturity (Table 5.3) following a calculation presented by Horsfield et al. (2006). The CO_2 respiration rates in mg per litre per year (Table 5.2) were converted into the required formate, acetate and oxalate feedstock in mg/g sediment per year using a stoichiometric conversion factor from CO_2 to the corresponding LMW organic acid and an approximate sediment density of 2 kg/l. Afterwards these data were compared to the detected formate, acetate and oxalate amounts released during the ester cleavage experiments corrected for the average amounts of these compounds at maturity levels higher than 0.55% R_0. At higher maturity levels, the decrease flattens representing an amount of LMW organic acids which is not released during diagenetic and early catagenetic transformation of the organic matter (non released residue). The minimum feedstock potential was calculated from the maximum respiration rate and vice versa. Depending on minimum and maximum respiration rates, or-

5.4. Results and discussion

	Minimum	Maximum
CO_2 respiration rates[1]		
mmol/l/a	$1 * 10^{-6}$	$1 * 10^{-3}$
mg/l/a	$4.4 * 10^{-5}$	$4.4 * 10^{-2}$
stoichiometric factor for conversion of LMWOA to CO_2		
formic acid (M = 46.03)		1.05
acetic acid (M = 60.05)		1.36
oxalic acid (M = 90.04)[2]		1.02
feedstock [mg/l/a]		
formate	$4.62 * 10^{-5}$	$4.62 * 10^{-2}$
acetate	$5.98 * 10^{-5}$	$5.98 * 10^{-2}$
oxalate	$4.49 * 10^{-5}$	$4.49 * 10^{-2}$
av. density of sediment [kg/l]	approx. 2	
feedstock [mg/g/a]		
formate	$2.31 * 10^{-8}$	$2.31 * 10^{-5}$
acetate	$2.99 * 10^{-8}$	$2.99 * 10^{-5}$
oxalate	$2.24 * 10^{-8}$	$2.24 * 10^{-5}$

Table 5.2.: *Assessment of required formate, acetate and oxalate respiration rates to sustain deep microbial life (modified after Horsfield et al., 2006).* [1]*CO_2 respiration rates from D'Hondt et al. (2002)* [2]*one oxalate provides two CO_2.*

5. Ester Bound Low Molecular Weight Organic Acids Linked To The Kerogen

ganic acid species detected and maturity level of the investigated coals, the feedstock potential of the NZ coals is highly variable. It ranges from several tens of thousand years to several hundred of million years (Table 5.3).

Because of the differences in the organo facies of the coals and the approximate age classifications, generation rates for small organic acids from the NZ coals are difficult to determine. Therefore, such calculations should be regarded as a first estimate only. Between sample G001978 of Oligocene to Miocene age (ca. 23 Ma; R_0 0.28%) and sample G001995 of approximately Middle Eocene age (ca. 43 Ma; R_0 0.52%) about 12.2 mg/g sediment formate and 2.4 mg/g sediment acetate was released during 20 Ma. For oxalate, the generation rate was calculated between sample G004544 of Miocene to Pliocene age (ca. 5 Ma; R_0 0.29%) and G001995 (see above). According to this, 1.4 mg/g sediment oxalate was released during ca. 38 Ma. This corresponds to a formate generation rate of $6.1 \cdot 10^{-7}$ mg/g sediment per year, for acetate $1.2 \cdot 10^{-7}$ mg/g sediment per year and for oxalate $3.7 \cdot 10^{-8}$ mg/g sediment per year, falling into the range of deep biosphere utilisation rates (Table 5.2), although at the lower end. However, metabolic rates in the deep biosphere are assumed to be on a very low level (Parkes et al., 2000). Thus, these approximate estimations show that during diagenetic alteration, the organic carbon rich formations can provide, indeed, a potential substrate feedstock, being large enough to sustain deep microbial life over geological time spans.

5.4. Results and discussion

	Eastern Southland		Waikato				West Coast		
Sample No.	G001978	G004541	G004544	G001982	G001983	G001980	G001996	G001995
R_0	0.28	0.29	0.29	0.40	0.41	0.44	0.52	0.52
Available acids				[mg/g sediment]				
Formate	14.3	5.7	8.6	10.7	3.9	2.3	2.5	2.1
Acetate	2.8	1.2	0.6	2.6	1.0	0.8	0.5	0.4
Oxalate	-	1.2	1.6	-	1.0	0.9	0.4	0.2
Estimated feedstock potential			[min. value in thousand years - max. value in million years]					
Formate	619	257	372	463	169	100	108	91
Acetate	94	40	20	87	33	27	17	13
Oxalate	-	54	71	-	45	41	18	9

Table 5.3.: *Calculated feedstock potential of the New Zealand coals investigated. Minimum values are calculated from maximum respiration rate, maximum values are calculated from minimum respiration rate of each compound. "Available acids" are calculated from the detected acids (Table 5.2) subtracted by an average residue amount, which appears not to be released during diagenesis. This "average non-released residue" was estimated from coal samples with $R_0 > 0.55\%$: for formate 2.2 mg/g sediment, for acetate 0.6 mg/g sediment and for oxalate 0.6 mg/g sediment.*

139

5. Ester Bound Low Molecular Weight Organic Acids Linked To The Kerogen

5.4.4. Mechanisms for the release of LMW organic acids from the coal matrix

In contrast to free organic compounds being part of the bitumen fraction and being partly dissolved in the pore water, the ester linked LMW organic acids investigated here are not directly available to act as a feedstock for deep microbial life. However, the presented data show that these compounds are gradually released during increasing maturation. Thus, there must be processes activating this part of the organic matter. Although we can only speculate on these processes, both abiotic as well as biotic processes are conceivable.

Temperature is most likely one of the crucial factors. In the upper sediments, buried organic matter passes through intense biodegradation processes. It is thought, that these processes initially lead to an increase of the recalcitrant proportion of the organic matter (Wellsbury et al., 1997; Parkes et al., 2007). However, with increasing burial depth, ambient temperature increases (on average about 30°C/km; (Allan and Allen, 1998)) affecting the stability of chemical bonds. The thermal supply of energy to the sedimentary system with depth increases the probability that the required activation energy for the cleavage of chemical bonds is achieved (Quingley and Mackenzie, 1988). During diagenesis, this might support abiotic bond cleavage of potential substrates e.g. by hydrolysis (*cf.* Chapter 9) and, thus, gradually activate the buried organic matter increasing its bioavailability. These thermally supported processes might therefore stimulate deep microbial activity (Parkes et al., 2007).

Parkes et al. (2007) simulated thermally supported alteration in long term laboratory experiments and showed continuously stimulated bacterial activity, indicating temperature induced activation of recalcitrant organic matter. Investigating coastal sediments using temperature heating experiments, Wellsbury et al. (1997) showed that the bioavailability of buried organic matter increases already at relatively low temperatures (30-40°C)

5.4. Results and discussion

indicated by enhanced release of acetate into the pore water. In coal environments the activation of recalcitrant organic matter by increasing geothermal heat with increasing burial depth might additionally be supported by catalytic effects in the pore water system. Lignites and coals contain high amounts of humic and fulvic acids. The kerogen itself bears free carboxylic acid groups and acidic phenol hydroxy groups. Furthermore, oxidation of pyrite in coals can cause low pH values of pore waters, an effect known from weathering of coals in coal mines, leading to the so called acid mine drainage (AMD), a problem for surface ecosystems located near coal mines (Olson et al., 1979; Christensen et al., 1996; Ludwig and Balkenhol, 2001; Blodau, 2006). Therefore, the pore water in coal seams often represents an acidic environment. In the presence of water and catalytic amounts of acids, esters form a chemical equilibrium with the hydrolysed organic acids and alcohols. The hydrolysed acids and alcohols can be removed by pore water flow or might simply be consumed by microorganisms. Therefore, acidic conditions in the pore water system are supposed to support a constant substrate release from coal matrices.

From early catagenesis to the main catagenetic phase, the relevance of thermogenic bond-cracking continuously increases. For example, acetate was found to be associated with thermogenic alteration of buried organic matter as a process of the formation of fossil fuel and represents a principal energy source for the deep production of bacterial methane (Cooles et al., 1987; Strapoc et al., 2008). However, thermogenic processes might not only support the release of ester-bond LMW organic acids as acids but also as CO_2 or in their reduced forms like methane and higher hydrocarbon gases.

Furthermore, it was shown that microorganisms are able to support the degradation of organic matter using extracellular enzymes (Arnosti et al., 1998; Arnosti and Jørgensen, 2003). Although such processes were so far only observed in marine or lacustrine sediments during early diagenesis, it is conceivable that this is another appropriate way for microorganisms to activate potential substrate sources in deep terrestrial systems as well.

5.5. Conclusion

In the current paper, a novel analytical approach was developed to receive low molecular weight compounds from alkaline ester cleavage reaction of coal samples of different maturity to assess the feedstock potential of these coals for deep microbial life. Test experiments on the applicability of the new method show best results in terms of recovery of low molecular weight (LMW) organic acids and high molecular weight (HMW) fatty acids as well as analytical handling when using methanol as the reaction solvent under reflux or in a closed vial in an oven.

Twelve coal samples showing a maturity range from lignite to high volatile bituminous coals taken from different locations on the North and South Island of New Zealand were investigated for ester linked LMW organic acids forming a potential substrate feedstock for deep microbial life. In almost all samples, significant amounts of formate, acetate and oxalate were detected, whereas the amounts releasable from the coal matrix decrease with increasing maturity. The main decrease of LMW organic acids is associated with the diagenetic phase reflecting structural alteration processes characterised by the loss of oxygen containing compounds. Concomitantly, this phase coincides with temperature condition where life in the deep subsurface is assumed to be still possible. First estimations of the feedstock potential and the generation rates of LMW organic acids indicate that the investigated NZ coals contain a substrate feedstock being large enough to sustain deep terrestrial microbial life over geological time spans.

5.6. Aknowledgements

We thank R. Sykes (GNS Science, New Zealand) for the selection of coal samples and providing valuable information about the selected sample material. Furthermore, we are grateful to Dr. P. Schaeffer (University Louis Pasteur in

5.6. Aknowledgements

Strasbourg, France) for intense and fruitful discussions about ester cleavage procedures and to Dr. S. Shouten (Royal Netherlands Institute for Sea Research (NIOZ), The Netherlands) and to two anonymous reviewers for their constructive and helpful comments. Kristin Günther and Cornelia Karger are thanked for IC and GC-MS measurement. Finally, we would like to thank the Helmholtz Association for financial support of this study.

6. Ester-Bound Fatty Acids And Alcohols Linked To The Kerogen

This chapter (except 6.7) was published in:

Glombitza, C.; Mangelsdorf, K.; Horsfield, B. (2009). Maturation related changes in the distribution of ester-bound fatty acids and alcohols in a coal series from the New Zealand Coal Band covering diagenetic to catagenetic coalification levels. Organic Geochemistry, 40(10):1063-1073.

6.1. Introduction

Wherever geochemists investigate the composition of sedimentary rocks, they observe that the biosphere on Earth has left its traces in the form of diverse organic biomass in the sediments. Although, in general, the main part of the decaying organic matter is recycled by microorganisms in the upper few surface centimetres to metres of sediments (Given et al., 1984; Hedges et al., 1985; Collins et al., 1992), continuously a significant proportion of it is incorporated into the sedimentary system. The buried organic matter is still exposed to various degradation and alteration processes, leading to the formation of complex macromolecular organic matter, the kerogen (Hedges, 1978; Durand, 1980; Rubinsztain et al., 1984; Tegelaar et al., 1989). Ongoing alteration processes mediated by microorganisms, temperature, pressure and time lead to a permanent compositional modification of the complex geopolymers with increasing maturity (Derbyshire et al., 1989; Rullköt-

6. Ester-Bound Fatty Acids And Alcohols Linked To The Kerogen

ter and Michaelis, 1990; Petsch et al., 2000; Vandenbroucke and Largeau, 2007). During diagenetic and early catagenetic alteration, these processes are accompanied by a significant loss of organic matter moieties containing various functional groups (e.g. carbonyl groups, hydroxy groups, sulphide groups), increasing the relative proportion of carbon in the residual organic macromolecules (Béhar and Vandenbroucke, 1987; Sinninghe-Damsté et al., 1989; Eglinton et al., 1994; Putschew et al., 1998; Vandenbroucke and Largeau, 2007; Petersen et al., 2008). Oxygen/carbon (O/C) ratios, determined by Rock-Eval pyrolysis, indicate (particularly for terrigenous organic matter) that the kerogen loses a significant proportion of oxygen-containing compounds especially during diagenesis and early catagenesis (Tissot and Welte, 1984a; Vu, 2008).

To investigate such structural alteration processes in the case of terrigenous organic matter on a compositional basis, selected coals from the New Zealand Coal Band provide an excellent sample set (Norgate et al., 1999; Vu, 2008). Coal-bearing formations located at the North and the South Island of New Zealand exhibit a series of coals of almost continuous maturity from Cretaceous to Tertiary age covering a maturity range from 0.25-3.0% vitrinite reflectance (R_0) (Suggate, 2000; Killops et al., 2002; Sykes and Snowdon, 2002). Thus, the coals of the New Zealand Coal Band reflect a wide range of maturities from peat to anthracite. The selected samples for the current study are from low to moderate maturity (R_0 0.27-0.80%), ranging from lignite to high volatile bituminous coals.

Pyrolysis experiments and chemical degradation procedures are appropriate tools to achieve a deeper insight into the compositional and structural changes of the complex organic matter during increasing maturity (Wollenweber et al., 2006). During pyrolysis, the macromolecular structure is thermally cracked and the produced volatile fragments, analysed by gas chromatography (Larter and Douglas, 1982; Horsfield, 1989; Larter and Horsfield, 1993), provide some valuable information about structural units of the source geopolymer. However, due to the radical thermal cracking process during pyrolysis, the information content about the different types of chemical bonds in the geopolymer is restricted. In this context, selective chemical degradation reactions provide a powerful approach to obtain more detailed information on chemical bonding structures in the complex organic coal macromolecules. Analytical procedures for the cleavage of the most common chemical bonds in kerogen such as oxygen (Schaeffer et al., 1995; Koopmans et al., 1997; Höld et al., 1998; Putschew et al., 1998; Schaeffer-Reiss et al., 1998; Wollenweber et al., 2006) and sulphur (Sinninghe Damsté et al., 1989a; Sinninghe-Damsté et al., 1989; Schouten et al., 1993) bonds are well established.

In a previous study, we already investigated the loss of low molecular weight (LMW) organic acids (formate, acetate and oxalate) forming a potential feedstock for deep microbial life from the complex organic matrix of coals from the New Zealand Coal Band with ongoing maturity (Glombitza et al., 2009b). In the current study, we now report on the release of high molecular weight (HMW) n-fatty acids and n-alcohols from the macromolecular New Zealand coal matrix during diagenetic to moderate catagenetic coalification levels.

6.2. Sample material

The investigated sample material was gathered in 2002 mainly from freshly exposed coal faces in mines or natural outcrops on the North and South Island of New Zealand and is part of the New Zealand Coal Band. The New Zealand Coal Band represents one of the best coal series in the world to investigate maturity-related processes in terrigenous kerogen. In the current study, 16 coal samples were examined covering diagenetic to moderate catagenetic coalification levels from 0.27-0.80% vitrinite reflectance (R_0) (Table 6.1).

6. Ester-Bound Fatty Acids And Alcohols Linked To The Kerogen

Sample	R_0 (%)	Basin	Formation	TOC	Fatty acids (µg/gTOC)					CPI_{FA}		Alc. (µg/gTOC)
					C_6-C_{12}	C_{13}-C_{19}	C_{20}-C_{30}	total		(C_{14}-C_{18})	(C_{20}-C_{30})	total
G001986	0.27	Northland	unknown	44.86	21	467	381	868		33.0	10.6	4011
G001988	0.27	Northland	unknown	45.10	432	623	369	1423		26.7	7.8	2815
G001978	0.28	Eastern Southland	GLM	52.44	36	436	130	602		12.6	4.8	1250
G001976	0.29	Eastern Southland	GLM	50.43	52	138	168	358		7.1	7.6	1485
G001982	0.40	Waikato	WCM	54.81	48	469	271	788		10.7	2.1	45
G001983	0.41	Waikato	WCM	53.82	63	157	209	428		5.1	2.6	54
G001980	0.44	Waikato	WCM	61.16	10	112	265	387		6.7	3.6	7
G001981	0.45	Waikato	WCM	60.75	21	107	322	430		3.8	3.0	n.d.
G001984	0.45	Waikato	WCM	63.33	9	97	257	342		3.3	2.3	8
G001992	0.49	Waikato	WCM	60.16	14	148	389	525		4.8	2.3	25
G001995	0.52	West Coast	BCM	67.76	15	578	178	771		17.9	2.3	134
G001996	0.52	West Coast	BCM	67.85	18	822	210	1049		23.9	2.1	58
G001994	0.61	Taranaki	Rakopi	63.99	16	941	223	1180		25.0	1.5	232
G001989	0.69	West Coast	BCM	73.33	17	210	55	281		14.0	1.1	15
G001993	0.76	West Coast	BCM	77.73	35	132	99	265		4.9	1.1	32
G001991	0.80	Taranaki	Mangahewa	73.83	18	79	41	137		4.6	1.2	200

Table 6.1.: *Information about the sample maturities (vitrinite reflectance R_0), sedimentary basins, lithological formations, total organic carbon (TOC) contents, ester-bound fatty acid and alcohol contents and carbon preference indices (CPI_{FA}) of short (C_{14}-C_{18}) and long (C_{20}-C_{30}) chain fatty acids of the New Zealand coal samples. Sample information, total organic carbon (TOC) values and R_0 data are provided by R. Sykes (GNS Science, New Zealand). Alc. = Alcohols, n.d. = not determined, GLM = Gore Lignite Measures, WCM = Waikato Coal Measures, BCM = Brunner Coal Measures*

The two samples from the Northland Basin and the two samples from the Eastern Southland Basin are in the lignite range, the six samples from the Waikato Basin are in the bituminous rank range and the four samples from the West Coast Basin as well as the two samples from the Taranaki Basin are in the rank range of high volatile bituminous coals. The Northland and Eastern Southland Basin samples represent diagenetic maturation levels, the Waikato Basin samples represent the transition to the early catagenesis phase and the West Coast and Taranaki Basin samples are in the stage of early to moderate catagenesis. The selected sample set covers a time period from Late Cretaceous to Pleistocene age.

6.3. Methods

The alkaline cleavage reaction was performed according to a method described earlier by Höld et al. (1998). However, some modifications were made in order to combine this method with an analytical procedure for the detection of low molecular weight organic acids as described earlier in Glombitza et al. (2009b). Additionally, the ester cleavage products (n-fatty acids and n-alcohols) were separated on a KOH impregnated separation column.

6.3.1. Sample preparation / saponification

A scheme of the applied analytical procedure is shown in Figure 6.1. 2 g of the freeze dried, ground and pre-extracted coal sample (to remove the bitumen fraction; see Glombitza et al. (2009b)) were suspended in 20 ml of 1 N KOH in methanol and refluxed for 5 h. After cooling to room temperature, 50 μg of erucic acid and 50 μg of 5α-androstan-17-one (both dissolved in methanol) were added as internal standards for compound quantification. The suspension was filtered and the residue was washed three times with 50 ml of methanol. The combined filtrates were split into two proportions,

6. Ester-Bound Fatty Acids And Alcohols Linked To The Kerogen

one being used for analysis of low molecular weight organic acids as reported in Glombitza et al. (2009b) and the other for the analysis of high molecular weight n-fatty acids and n-alcohols. The solvent (methanol) was removed under reduced pressure and an alkaline slurry remained containing the fatty acids in the form of their corresponding potassium salts and the non volatile HMW alcohols. The remaining alkaline slurry was dissolved in ca. 20 ml water and acidified with diluted HCl (5%) to a pH of 3. This solution was extracted three times with 50 ml dichloromethane in a separation funnel. The combined dichloromethane extracts were dried with sodium sulphate before the solvent was finally removed.

Separation of fatty acids and alcohols was achieved by column chromatography using alkaline-impregnated silica gel as stationary phase and a slight overpressure (ca. 3 bars) of nitrogen. To prepare the stationary phase, 5 g silica gel (Merck, 60) were suspended in a mixture of 20 ml 0.5 N isopropanolic KOH solution and 20 ml dichloromethane using ultra sonication for 20 min. The slurry was filled into a column. Subsequently, the packed column was cleaned and excess of KOH was removed with approximately 100 ml of dichloromethane before adding the sample extract onto the column. Fatty acids were precipitating in form of their potassium salts on the alkaline silica gel, while the neutral polar fraction containing the alcohols was eluted with 200 ml of dichloromethane. Afterwards, the potassium salts were transformed back to the free fatty acids by elution with 200 ml of 2% formic acid in dichloromethane, resulting into the fatty acid fraction. Finally, after derivatization (see below), both fractions were analysed by gas chromatography-mass spectrometry.

Additionally, the filtration residues before and after the alkaline cleavage experiment were investigated by open system pyrolysis-gas chromatography. The weight of the residue directly achieved after filtration of the reaction mixture from alkaline cleavage reaction was found to be approximately 25% heavier than the initial pre-extracted coal samples due to adsorption of KOH on the coal surface. When dispersed in water or methanol, the alkaline

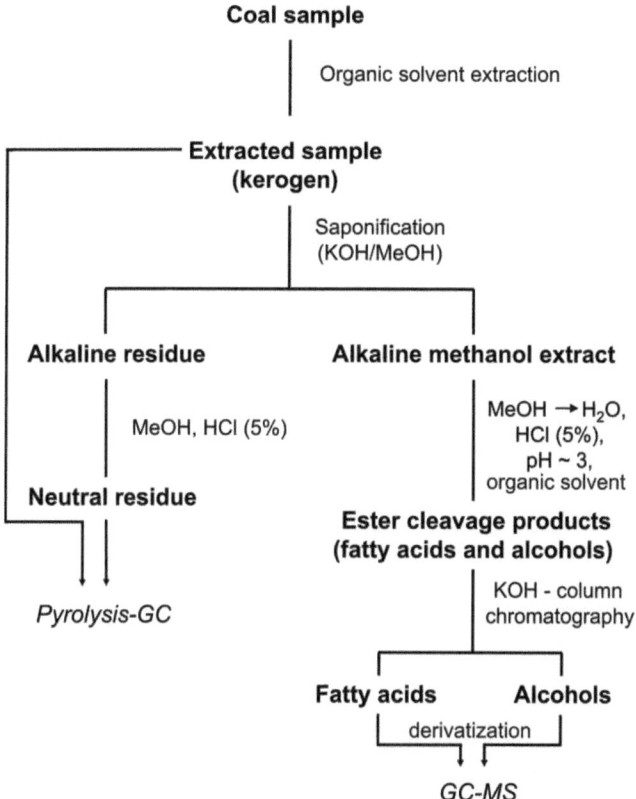

Figure 6.1.: *Separation scheme outlining the analysis of ester-bound fatty acids and alcohols from coal samples using GC-MS and open system pyrolysis-GC.*

cleavage residue samples show high pH values around pH 12. In contrast to alkaline ester cleavage reactions performed with kerogen type II samples (Höld et al., 1998; Schaeffer-Reiss et al., 1998), the surface adsorbed KOH appears to be not removable from the coal matrix by simple washing with methanol or methanol/water mixtures. To remove the adsorbed KOH, the samples were dispersed in methanol and diluted HCl (5%) was added drop wise to the stirred suspension for neutralisation. The pH values were controlled by a glass electrode (Schott). This solution was stirred for ca. 10 min until the pH had reached a value of approximately 3. The residue was filtered and washed with MeOH until the pH of the extract became neutral. Neutralisation of the cleavage residue prior to open system pyrolysis was necessary, because the excess of KOH negatively affected the amounts of pyrolysis products, especially in case of phenolic compounds (data not shown).

6.3.2. Gas chromatography-mass spectrometry (GC-MS)

Prior to GC-MS analysis, the fatty acids were derivatised with diazomethane to obtain fatty acid methyl esters. The alcohol fractions were treated with N-methyl-N-(trimethylsilyl)trifluoroacetamide (MSTFA) to obtain the corresponding trimethylsilyl(TMS)-ethers.

GC-MS analysis was performed using a Trace GC Ultra (Thermo Electron Corporation) linked to a Dual Storage Quadrupole (DSQ) mass spectrometer (Thermo Electron Corporation). The GC oven temperature was programmed from 50°C (1 min holding time) to 370°C with a heating rate of 3°C/min and was held at 370°C for 5 min. Than, the temperature was reduced to 350°C with a rate of 15°C/min and held at 350°C for 20 min. The carrier gas was helium with a constant flow of 1 ml/min. Sample volume injection of 1 μl was performed with a PTV Splitless Injector programmed from 50-300°C at 10°C/s and held at 300°C for 10 min. The GC capillary column used was a BPX5, length 50 m, diameter 0.22 mm and 0.25 μm film thickness. In the coupled mass spectrometer ions were obtained by electron impact (EI) ionisation at 70 eV at a source temperature of 230°C. The scan interval was 50-650 amu with 2.5 scans/sec. Compound signals were integrated using the software Xcalibur™ 1.4 SR1 (Thermo Electron Corporation).

6.3.3. Open system pyrolysis GC

For open system pyrolysis, dried coal samples before and after alkaline treatment were used. 2-5 mg of the fine powdered sample was weighed into a glass capillary tube and fixed using thermally pre-cleaned quartz wool. The capillary was placed into a pyrolysis oven linked to a gas chromatograph and heated to 300°C. Products released up to this temperature were vented in order to remove volatile compounds. Afterwards, the system was heated from 300°C to 650°C using a heating rate of 49°C/min and the pyrolysis products were collected in a liquid nitrogen cooling trap. After the pyrolysis was completed, the collected products were released by heating the cold trap to 300°C and introduced into an Agilent 6890A gas chromatograph equipped with a flame ionisation detector (FID). The used carrier gas was Helium. The GC capillary column used was a HP-Ultra 1, length 50 m, diameter 0.32 mm and 0.52 μm film thickness. n-Butane was used as standard to quantify the generated products. The identification of the pyrolysis products was based on reference chromatograms. The used software was Agilent GC ChemStation (Rev. A.09.03).

6.4. Results and discussion

6.4.1. Alcohols and fatty acids cleaved from NZ coal samples

Figure 6.2 shows typical GC-MS chromatograms of alcohol and fatty acid fractions of a low-rank (G001978; Eastern Southland Basin, R_0: 0.28%) and a moderate-rank (G001993; West Coast Basin, R_0: 0.76%) sample. In addition to n-alcohols and n-fatty acids, occurring in almost all samples (Table 6.1), the cleavage products contain a series of other compounds such as ω-hydroxy fatty acids, isoprenoids, steroids and triterpenoids. However, these compounds appear to be strongly related to the different facies and will not be discussed here (*cf.* Additional Section 6.7).

6. Ester-Bound Fatty Acids And Alcohols Linked To The Kerogen

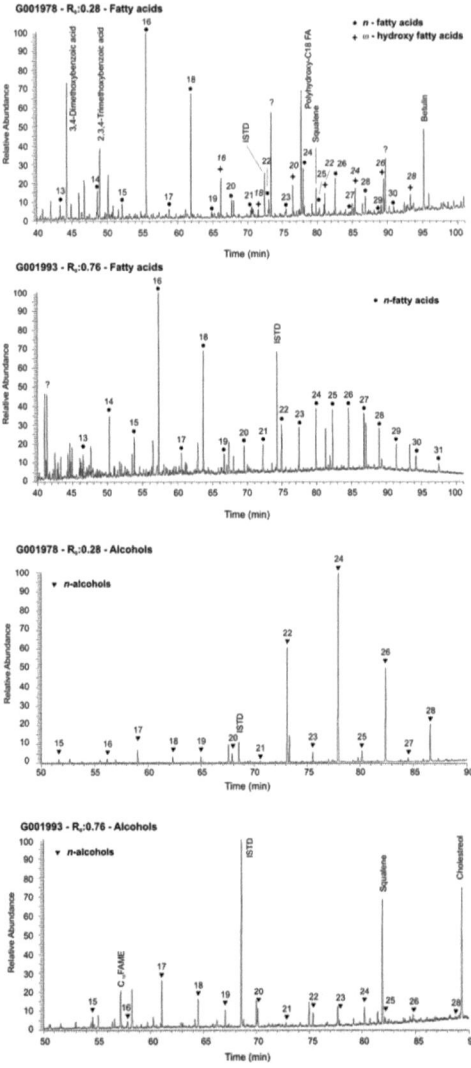

Figure 6.2.: *GC-MS-chromatograms of alcohol and fatty acid fractions of a low-rank (G001978) and of a moderate-rank (G001993) sample. ISTD = internal standard (fatty acid fraction: erucic acid; alcohol fraction: 5α-androstan-17-one); Numbers indicate carbon numbers of n-fatty acids or n-alcohols, respectively.*

6.4. Results and discussion

Highest concentrations of ester-bound alcohols (up to about 4 mg/gTOC, sample G001978) were detected in the very immature coal samples of the Northland Basin and the Eastern Southland Basin (vitrinite reflectance R_0<0.3%; Table 6.1, Figure 6.3a). Compared to the amounts of ester-bound fatty acids (Figure 6.3b) in the early diagenetic samples, the concentration of alcohols is up to 4.5 times higher. These high concentrations at low maturity levels may be due to the presence of a high number of benzoic kerogen core acids esterified with alcohol moieties. Terrigenous kerogen contains numerous aromatic units bearing functionalities like carboxylic acids (Hatcher, 1990; Vandenbroucke and Largeau, 2007; Petersen et al., 2008). The concentrations of released alcohols show a rapid decrease during diagenesis (R_0 0.27-0.4%) and remain at very low levels (about 44 μg/gTOC, e.g. sample G001982; Table 6.1 and Figure 6.3a) during the early and main catagenetic phase.

Fatty acids were also detected among the ester cleavage products of all samples investigated. Initially, the total amounts of fatty acids also rapidly decrease with increasing maturity of the coal samples (Figure 6.3b) from 1.42 mg/gTOC (sample G001988, R_0 0.27%) down to 0.34 mg/gTOC (sample G001984, R_0 0.45%). However, in contrast to the rapid and continuous decrease of the alcohols, the amounts of fatty acids show an increase during early catagenesis, before decreasing again to low values during the main catagenetic phase (0.14 mg/gTOC in sample G001991, R_0 0.8%).

A deeper insight into the distribution of the ester-bound alcohols and fatty acids in the selected coal samples at different levels of maturity is shown in Figure 6.4. During early diagenesis, the released alcohols show only a single main maximum located around C_{22} or C_{24} (Figures 6.4a-6.4b) and a strong even over odd carbon number predominance. The distribution of long chain alcohols indicates an origin from higher terrigenous plant debris (Eglinton and Hamilton, 1967; Kolattukudy, 1970; Oldenburg et al., 2000; Yamamoto et al., 2008). The different alcohol distribution patterns obtained from the immature samples of the Northland Basin (Figure 6.4a) and

6. Ester-Bound Fatty Acids And Alcohols Linked To The Kerogen

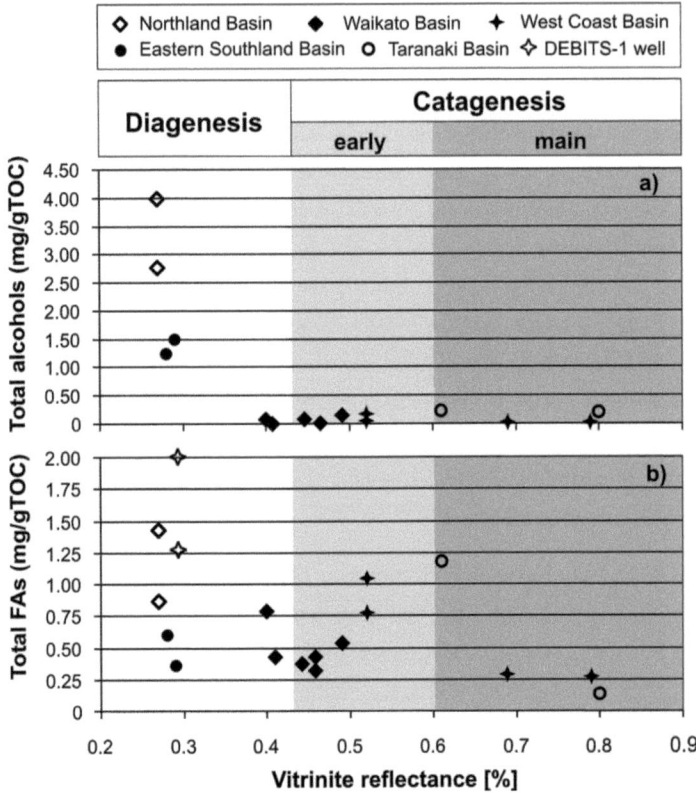

Figure 6.3.: *Concentrations (in mg/gTOC) of total alcohols (a) and total fatty acids (b) released by alkaline ester cleavage from a series of New Zealand coal samples of different maturity. (The two DEBITS-1 well samples were added to the fatty acid data for comparison with the free fatty acids from the bitumen fraction (cf. Chapter 7), data can be found in Table 8.2.)*

Eastern Southland Basin (Figure 6.4b) are characteristic for these sedimentary basins and indicate differences in the organofacies and therefore in the contributing plant communities. Furthermore, Figures 6.4a-6.4e show that the huge amounts of ester-linked alcohols, present in the immature samples, have already disappeared during late diagenesis (Figures 6.4c-6.4e;

6.4. Results and discussion

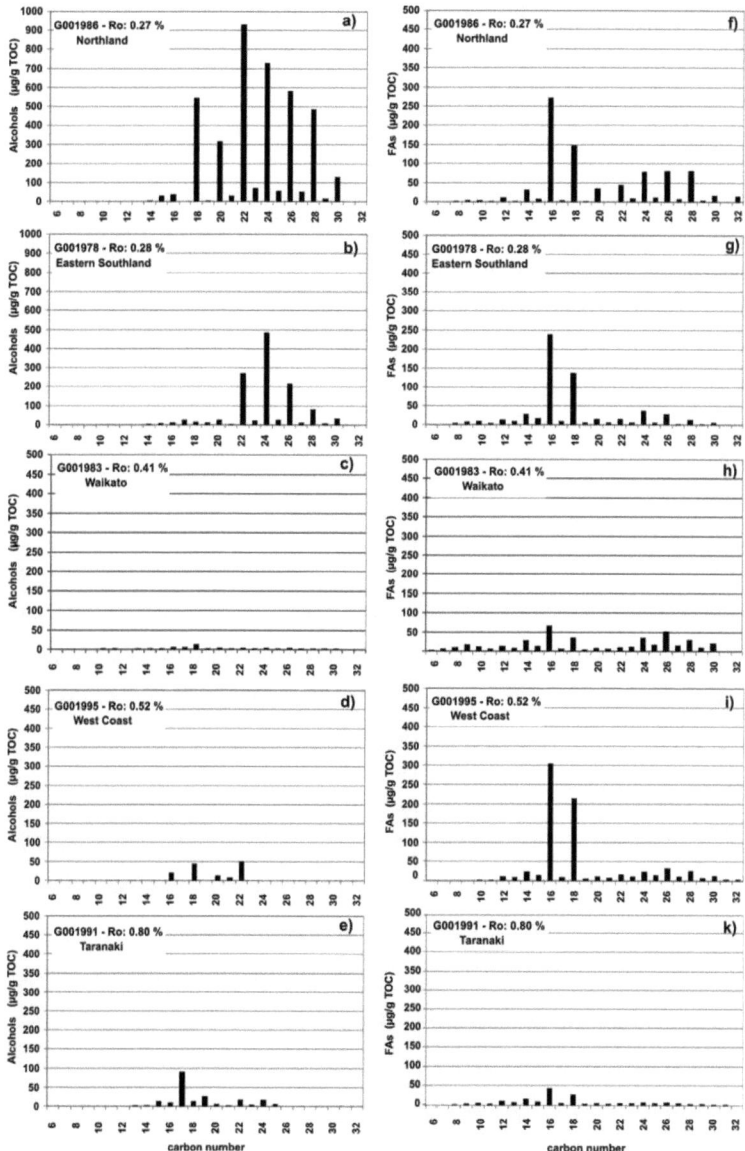

Figure 6.4.: *Alcohol (a-e) and fatty acid (FA; f-k) distribution patterns of selected coal samples from early to moderate coalification level (0.27% to 0.80% R_0).*

note the different scales of the y-axis between Figures 6.4a-6.4b and 6.4c-6.4k). During early and main catagenesis, only very low amounts of ester-cleaved alcohols in the range of C_{13}-C_{25} can be observed (Figures 6.4c-6.4e).

In contrast to the alcohols, the fatty acids show a distinct bimodal distribution pattern (Figures 6.4f-6.4k) with a first maximum at C_{16} and C_{18} fatty acids, as well as a second one centred around C_{24} and C_{26}. In terrigenous environments, short chain fatty acids might have many different sources since they are common constituents in eukaryotes (Cranwell, 1974; Shameel, 1990; Oldenburg et al., 2000) and might be incorporated during early diagenesis into the kerogen matrix. However, C_{16} and C_{18} fatty acids are also the main fatty acids in bacterial cell membrane phospholipids (White et al., 1979; Harvey et al., 1986; Zink et al., 2003) and, therefore, bacterial biomass provides another potential source for these biomarkers. Long chain fatty acids in the current predominance of even carbon numbered fatty acids are typical for terrigenous higher land plant material (Kolattukudy, 1966; Eglinton and Hamilton, 1967; Kvenvolden, 1967; Franke and Schreiber, 2007; Yamamoto et al., 2008). Regarding the terrigenous-derived long chain fatty acids, a continuous decrease in concentration can be observed with increasing maturity of the organic matter. In contrast, for the short chain fatty acids, such a clear decrease is not obvious. During diagenesis, an initial decrease is observed (R_0: 0.27-0.41%, Figures 6.4f-6.4h), followed by an intermittent increase (R_0: 0.52%, Figure 6.4i) representing the early catagenetic phase, before decreasing again at higher maturity (R_0: 0.80%, Figure 4k).

These results indicate that the intermittent rise of the total ester-linked fatty acids during early catagenesis observed in Figure 6.3b is mainly due to the contribution of the short chain fatty acid proportion among the ester cleavage products. In Figures 6.5a and 6.5b, the total fatty acid maturity profile is separated into its long chain (C_{20}-C_{32}) and short chain (C_{14}-C_{19}) fatty acid proportions. The amounts of long chain fatty acid are lower than those of the short chain fatty acids. For the long chain fatty acids, a slow,

6.4. Results and discussion

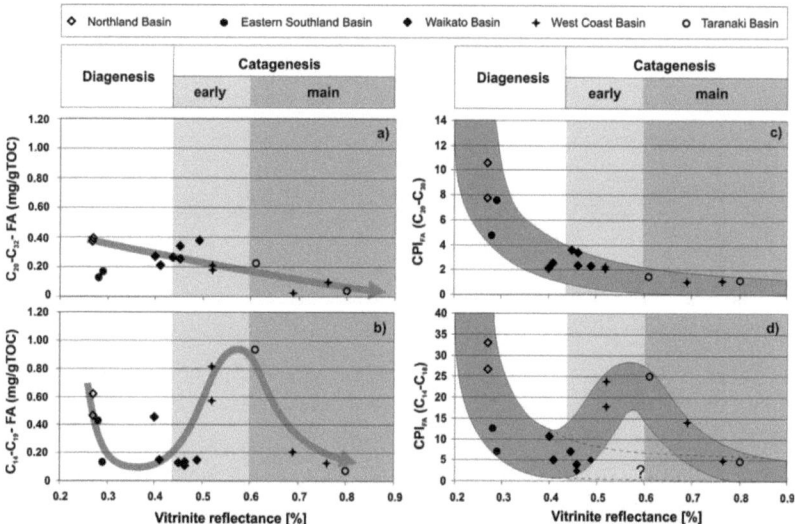

Figure 6.5.: *Concentrations (in mg/g TOC) of long chain fatty acids (a) and short chain fatty acids (b) released by alkaline ester cleavage from a series of New Zealand coal samples of different maturity. Carbon preference index (CPI_{FA}) for long chain fatty acids (C_{20}-C_{30}) (c) and short chain fatty acids (C_{14}-C_{18}) (d) versus sample maturity.*

but general, maturity-related decrease is visible (Figure 6.5a), whereas the short chain fatty acids reveal, after an initial decrease during diagenesis, an intermittent increase during early catagenesis with a subsequent decrease during the main catagenesis stage (Figure 6.5b). For comparison, the rank-related concentration of the kerogen-linked low molecular weight organic acids (LMWOA) formate, acetate and oxalate described in Glombitza et al. (2009b) show no intermittent rise during early catagenesis. The LMWOA patterns reveal a strong decrease during diagenesis, gradually slowing down during the early and main catagenesis.

When using the amount of the ester-linked fatty acids to monitor maturity dependent compositional changes in the kerogen matrix of different maturity, it has to be kept in mind that the abundance of fatty acids can be

6. Ester-Bound Fatty Acids And Alcohols Linked To The Kerogen

influenced by the initial organofacies forming the coal organic matter. Thus, some of the scattering around the general trend in Figures 6.5a and 6.5b might be due to facies-related differences of the organic source material.

6.4.2. Carbon preference index of short and long chain fatty acids

The different organofacies of the investigated coals from different sedimentary systems might to some extent obscure maturity-related trends in the coal series. Thus, to investigate compositional changes in geopolymers of different maturities, it would be most effective to use a facies-independent, but maturity-dependent, parameter. For the fatty acids, such a parameter appears to be provided by the ratio of even to odd carbon numbered fatty acids. In immature organic matter, fatty acids show a high even over odd carbon number predominance according to the fatty acid biosynthesis from acetyl (C_2) units (Kolattukudy, 1966; Franke and Schreiber, 2007; Yu et al., 2008). Kvenvolden (1967) summarized a series of studies on fatty acids obtained from different sediments covering a maturity range from Precambrian to recent sediments. These studies show that initially, in recent sediments, the even fatty acids are predominating over the odd fatty acids, but that in ancient sediments and in crude oils, this predominance decreases (Cooper, 1962; Kvenvolden, 1966). Cranwell (1974) uses the even over odd carbon number predominance for fatty acids obtained from lake sediments to reveal their terrigenous origin by comparing the carbon number predominance found in woodland humus and surface peat samples. In the same study, it was also shown that the even over odd predominance of fatty acids in the older peat sample is lower than in the surface peat sample. High even over odd carbon number predominance can also be observed for the fatty acids of the investigated immature coal samples shown in Figures 6.4f and 6.4g. Long chain fatty acids being derived from terrigenous plant communities can only be introduced into the sediments during the early sedimentation stage,

6.4. Results and discussion

and then, undergo continuous alteration. In Figures 6.4f to 6.4k, it can be observed that the even over odd carbon number ratio of the terrigenous long chain fatty acids clearly decreases with increasing maturity, indicating that the carbon number predominance vanishes with ongoing alteration of the coal matrix.

Based on these results and in order to reduce variations in the ester-bound fatty acid concentrations due to differences in the organofacies, we applied a carbon preference index for fatty acids (CPI$_{FA}$) to the New Zealand coal series. This parameter is, to some extant, analogue to the odd over even carbon number predominance index (CPI) for n-alkanes published by Bray and Evans (1961). However, the CPI$_{FA}$ relates the amounts of the predominating even carbon numbered fatty acids to the amounts of the neighbouring odd carbon numbered fatty acids. For the calculation of the CPI$_{FA}$ of the short chain fatty acids (equation 6.1), the amounts of C_{13} to C_{19} fatty acids were used. For the calculation of the CPI$_{FA}$ of the long chain fatty acids (equation 6.2), the amounts of C_{19}-C_{31} fatty acids were taken into account.

$$CPI_{FA}(C_{14} - C_{18}) = \frac{1}{2} \left(\frac{\sum_{n=7}^{9} C_{2n}}{\sum_{n=7}^{9} C_{2n-1}} + \frac{\sum_{n=7}^{9} C_{2n}}{\sum_{n=7}^{9} C_{2n+1}} \right) \qquad (6.1)$$

$$CPI_{FA}(C_{20} - C_{30}) = \frac{1}{2} \left(\frac{\sum_{n=10}^{15} C_{2n}}{\sum_{n=10}^{15} C_{2n-1}} + \frac{\sum_{n=10}^{15} C_{2n}}{\sum_{n=10}^{15} C_{2n+1}} \right) \qquad (6.2)$$

The change in the fatty acid carbon number predominance with increasing maturity might be explained by different decomposition rates of even and odd carbon numbered fatty acids. The abiotic processes leading to the decomposition of esters in sedimentary organic matter are either chemical decomposition by hydrolysis in pore waters during diagenesis and presumably early catagenesis, or thermal decomposition during catagenesis. Thermal decomposition is a unimolecular reaction which follows a first order kinetic. In contrast, hydrolysis is a bimolecular reaction usually following a second order kinetic. However, due to the huge excess of pore water molecules relative to the amounts of ester functionalities, a pseudo first order kinetic can

6. Ester-Bound Fatty Acids And Alcohols Linked To The Kerogen

be assumed. Thus, in both cases, the decomposition of ester-bound fatty acids can be described by first order reaction rates (equation 6.3, ν: reaction rate, t: time, k: reaction rate constant, [ester]: concentration of ester within the kerogen, (Atkins, 1986)).

$$\nu = -\frac{d[ester]}{dt} = k[ester] \qquad (6.3)$$

It can be deduced from equation 6.3 that the reaction rate only depends on the concentration of the different fatty acid esters in the sedimentary organic matter. Therefore, the more abundant even carbon numbered fatty acid esters are decomposed with higher rates than the odd carbon numbered esters until their concentration reaches similar levels, and the CPI_{FA} approaches a value of one. It is important to note that the CPI_{FA} only expresses the level of maturity (or degree of alteration) of the organic matter and provides no information about the total amount of the fatty acids cleaved from the coal matrix. Therefore, a high CPI_{FA} represents immature fresh material and a low CPI_{FA} more mature and more altered material. Figures 6.5c and 6.5d show the calculated CPI_{FA} values plotted versus the vitrinite reflectance data for the long and short chain fatty acid distributions, respectively. For the long chain fatty acids (Figure 6.5c), a maturity-related exponential decrease can be observed during the diagenetic phase, showing a continuous maturation of the ester-bound fatty acid fraction and, therefore, approving that the CPI_{FA} reflects ongoing maturation. Even those samples with comparatively low (samples G001976 and G001978 from the Eastern Southland Basin) and high (samples G001981 and G001992 from the Waikato Basin) total amounts of fatty acids now join into the same evolutional trend outlined by the CPI_{FA} of the other samples (Figures 6.5a and 6.5c). Concomitantly, these data show that the even over odd carbon number predominance of the long chain fatty acids vanishes during diagenesis and early catagenesis. In the samples from the main catagenetic stage, the carbon number predominance is almost absent and the CPI_{FA} shows values close to unity.

6.4. Results and discussion

In general, the CPI$_{FA}$ values (Figure 6.5d) of the short chain fatty acids are higher than those of the long chain fatty acids. The CPI$_{FA}$ values initially also show an exponential decrease during diagenesis and the transition to the early catagenesis, indicating the ongoing maturation of the fatty acids and the loss of the even carbon number predominance with increasing maturity. Even sample G001982 (R$_0$ 0.4%) from the Waikato Basin now plots within the range of the other samples from the Waikato Basin, indicating that the high amounts of ester-bound fatty acid detected in this sample are due to facies variations. However, in contrast to the trend indicated by the long chain fatty acids, the samples from the West Coast Basin (samples G001995 and G001996) and Taranaki Basin (sample G001994) show unexpected high CPI$_{FA}$ values during the early catagenetic stage. This interval was followed by a subsequent decrease during the phase of main catagenesis, until samples G001993 (R$_0$ 0.76%) and G001991 (R$_0$ 0.8%), both also from the West Coast and Taranaki Basin, show again CPI$_{FA}$ values being in the trend (dotted line) outlined by the long chain fatty acids and which might be expected for this advanced level of maturity. The early catagenesis samples with the high CPI$_{FA}$ values are the same samples showing very high amounts of ester-linked short chain fatty acids in Figure 6.5b.

6.4.3. Potential sources of short chain fatty acids during early catagenesis

The change from low to high amounts of short chain fatty acids during the early catagenetic stage coincides with the change from the Waikato Basin to the West Coast and Taranaki Basin samples (Figure 6.5b). On a first view, this suggests that different facies might be an explanation for these higher amounts of ester-linked short chain fatty acids. However, as mentioned above, these samples show also very high CPI$_{FA}$ values, being in the maturity range of early diagenesis (Figure 6.4d), when compared to the other samples from the New Zealand coal series. In contrast to this, the

6. Ester-Bound Fatty Acids And Alcohols Linked To The Kerogen

corresponding CPI_{FA} values of the long chain fatty acids show already an advanced maturity level fitting into the maturity trend outlined by the other coal samples. Thus, if facies variation is the reason for this observation, the short chain fatty acids appear to be less altered than the long chain fatty acids. This is a bit puzzling but might be the result of the decomposition of extremely different initial amounts of short and long chain fatty acids in the organic source material.

Another interesting and very intriguing explanation for these high amounts of short chain fatty acids with a high CPI_{FA} signature at this advanced maturity level is the supply of relatively immature organic material to the kerogen matrix. In such deep sedimentary systems, the only conceivable source of organic material with an immature signature are deep microbial communities. C_{16} and C_{18} fatty acids, being the main contributors to the short chain fatty acid maximum in the investigated samples (Figures 6.3f-6.3k), are, in their saturated but also unsaturated form, essential constituents of phospholipids forming bacterial cell membranes (White et al., 1979; Chappe et al., 1982; Ratledge and Wilkinson, 1988; Mangelsdorf et al., 2005). After cell death, this microbial biomass might be incorporated into the macromolecular matrix in the surrounding sediment, whereas the unsaturated fatty acids might lose their double bonds due to ongoing geochemical alteration processes.

If microbial biomass is the source of these short chain fatty acids, the question remains why this significant increase just occurred during the phase of early catagenesis. A reason for this might be that during this phase, the thermal degradation of the organic matter slowly commenced releasing hydrocarbons from the kerogen matrix into the surrounding sediments. These hydrocarbons form a potential feedstock for deep microbial life and might have stimulated microbial activity in this sedimentary section. Investigating thermally-generated hydrocarbons in the coal samples of the New Zealand Coal Band, Vu et al. (2009) were able to show that the onset of thermal degradation is located at a maturity level of about 0.43% R_0 (Sug-

6.4. Results and discussion

gate Rank S_r: 5-6; (Suggate, 2000)), characterised by a slow, but gradual, increasing generation of hydrocarbons. In the coals from the New Zealand Coal Band, this early catagenesis stage is followed by an intensive hydrocarbon release, the onset of intense oil generation, at about 0.6% R_0 (about S_r: 9). This phase of intensified hydrocarbon generation for the NZ coals was also shown by bitumen index data (BI=S1/TOC) obtained from Rock-Eval pyrolysis experiments (Sykes and Snowdon, 2002) and by the yields of thermal extracts obtained from thermal extraction- and pyrolysis-gas chromatography experiments (TE-Py-GC; (Sykes and Johansen, 2007)). Using kinetic modelling, in situ gas calibration, ^{14}C-radiolabelling of methanogenesis, bacterial cell counts and analysis of intact phospholipids (being indicators for microbial life; (White et al., 1979; Zink et al., 2003, 2008) in samples from ODP Leg 190 (Sites 1173, 1174 and 1177) in the Nankai Trough, offshore Japan, it was shown by Horsfield et al. (2006) that thermal degradation reactions of sedimentary organic matter appear in the same part of the sedimentary column as where a deep biosphere exists. In long term heating experiments (up to 500 days) using surface sediment slurries and sequential heating experiments, Parkes et al. (2007) were able to show that a temperature activation of recalcitrant organic matter is able to stimulate microbial activity, as indicated by a temperature-related enhanced production of CH_4 (ca. 30-50 °C), H_2 and acetate (ca. 50-80 °C).

It might be argued that the thermally-induced release of hydrocarbons from the kerogen matrix and the incorporation of the microbial biomass into the coal matrix contradict. However, it is conceivable that at early catagenetic maturation levels, the moderate thermal conditions causing a slow but gradual release of some compounds on the other hand might also support chemical reactions leading to the incorporation of e.g. dead microbial biomass from the pore waters into the kerogen matrix. However, at higher temperature conditions, the release of compounds continuously dominates as indicated by the intensive hydrocarbon generation beginning with the onset of the main catagenesis (approx. 0.6% R_0) as observed by Vu et al. (2009),

Killops et al. (1998) and Sykes and Snowdon (2002). From this maturation level, the decreasing CPI$_{FA}$ values indicate no additional supply of microbial biomass into the coal matrix (Figure 6.5d). Temperatures between 80 and 90 °C are suggested to form the upper temperature limit for deep subsurface microbial life (Connan, 1984; Wilhelms et al., 2001; Head et al., 2003). Using the kinetic model of Sweeney and Burnham (1990) with an average heating rate for geological systems of 2 °C per Ma, a temperature of about 100 °C is roughly accessed for the beginning of the main catagenesis with 0.6% R_0. Thus, temperature conditions appear to become incompatible with microbial life at the transition to the main catagenetic phase, which coincides with the decrease of CPI$_{FA}$ values of the short chain fatty acids from this maturity level.

6.4.4. Open system pyrolysis of the NZ coal samples

The selected coal samples from the NZ Coal Band were investigated to evaluate whether the release of fatty acids and alcohols by alkaline ester cleavage reaction can also be observed by open system pyrolysis. For this purpose, the pyrolysis products of solvent pre-extracted coal samples prior and after ester cleavage were compared.

Figures 6.6a and 6.6b show the alkene pyrolysis products of sample G001978 (early diagenesis, 0.28% R_0) and sample G001982 (late diagenesis, 0.4% R_0) before and after alkaline hydrolysis. In the immature untreated coal samples of the Eastern Southland Basin (sample G001978; Figure 6.6a), alkenes with a strong even over odd carbon number predominance ranging from C_{22} to C_{28} are observed. After alkaline treatment, the alkene concentrations are significantly depleted. Comparing the pyrolysis results with the alcohol distributions obtained after ester cleavage, it becomes obvious that the formerly ester-bound alcohols from the Eastern Southland Basin sample show also a strong even over odd carbon number predominance in the same range (sample G001978; Figure 6.6c). In Figure 6.3a,

6.4. Results and discussion

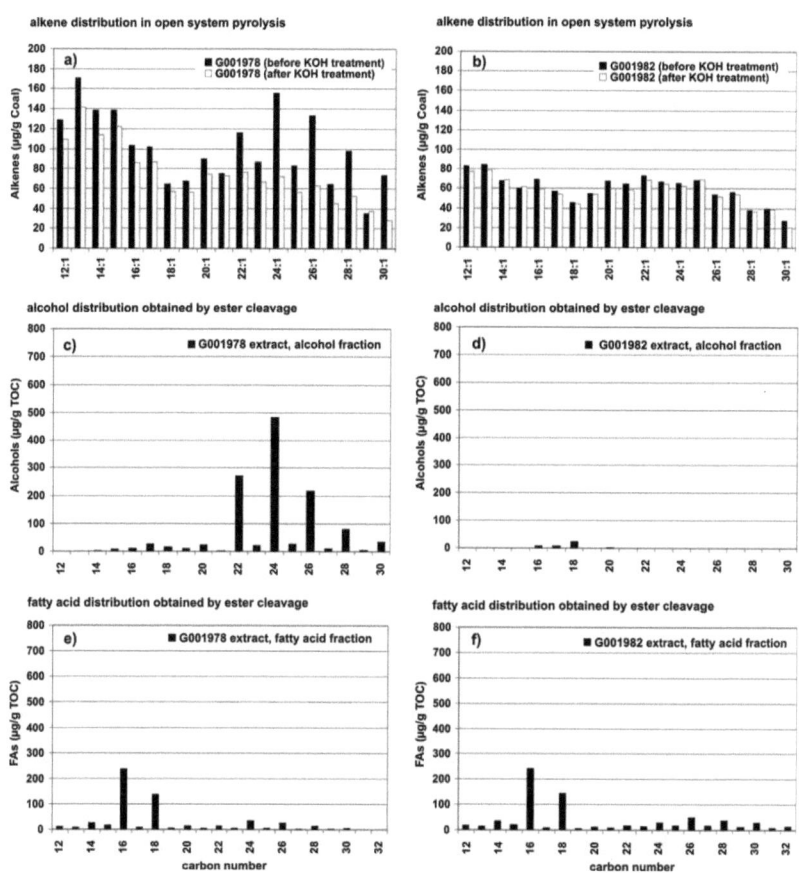

Figure 6.6.: *Distribution pattern of alkenes obtained by open system pyrolysis of sample G001978 (Eastern Southland Basin) from early diagenesis (R_0: 0.28%) (a) and of sample G001982 (Waikato Basin) from late diagenesis (R_0: 0.40%) (b) before and after alkaline ester cleavage. For comparison, the corresponding distribution patterns of ester-bound alcohols (c and d) and fatty acids (e and f) obtained after alkaline cleavage reaction are shown.*

it can be seen that ester-bound alcohols rapidly decrease already during the phase of early diagenetic alteration. Thus, in the more mature samples reaching late diagenesis, alcohols are found only in very low amounts

6. Ester-Bound Fatty Acids And Alcohols Linked To The Kerogen

(sample G001982; Figure 6.5d) and almost no changes in the alkene distribution after saponification was detected (sample G001982; Figure 6.5b). This suggests that in the low-rank samples, a significant proportion of the long chain alkenes, visible in open system pyrolysis, originally derives from ester-bound alcohols present in the kerogen matrix. Furthermore, it is notable that these changes in the alkene distribution after alkaline cleavage are not observed in the alkane distribution (data not shown). The alkaline-treated samples from the Eastern Southland Basin show the same alkane distribution as the untreated samples. Consequently, ester-bound alcohols appear to produce only alkenes as products in open system pyrolysis. The formation of alkenes from alcohols due to pyrolytic-induced ester scission was also reported by Asperger et al. (1999) investigating natural waxes and specific model ester compounds.

In contrast to the ester-bound alcohols (Figures 6.5c and 6.5d), the ester-linked fatty acids (Figures 6.5e and 6.5f) do not contribute significantly to the alkane or alkene distribution pattern obtained by open system pyrolysis in the coal samples e.g. via decarboxylation. This coincides also with the observations made by Asperger et al. (1999) demonstrating a higher thermal stability for the formerly ester-bound fatty acids during pyrolysis-GC.

6.5. Conclusion

A maturity series of 16 lignite and coal samples of low to moderate maturation levels (R_0: 0.27-0.80%) taken from the New Zealand Coal Band were investigated using alkaline ester cleavage experiments. The alcohols and fatty acids obtained after saponification were investigated using GC-MS.

The concentrations of the formerly ester-bound alcohols rapidly decrease during diagenesis and remain low during the early and main catagenesis stages. The ester-bound fatty acids initially also decrease, but show an intermittent increase during early catagenesis before decreasing again

6.5. Conclusion

during main catagenesis. This intermittent increase was found to be related only to the short chain fatty acids mainly with 16 and 18 carbon atoms.

To obtain a facies-independent and maturity-related signal, an even over odd carbon preference index for fatty acids (CPI_{FA}) was calculated. For the long chain fatty acids (C_{20}-C_{30}) representing terrigenous plant material, this parameter shows a constant and rapid decrease during diagenesis and early catagenesis and is close to the unity during the main catagenesis stage. This shows that the CPI_{FA}, at least for the long chain fatty acids, represents ongoing maturation of organic matter.

Similar to the concentration of the formerly ester-linked short chain fatty acids, dominated by the C_{16} and C_{18} fatty acids, their CPI_{FA} (C_{14}-C_{18}) shows also an initial rapid decrease during diagenesis and an intermittent increase during early catagenesis followed by a decrease during the main catagenesis. This suggests the contribution of a significant proportion of apparently less altered organic material during early catagenesis. A reason for this might be extremely different amounts of short and long chain fatty acids in the original source material causing this relatively high CPI_{FA} (C_{14}-C_{18}) values during early catagenesis. However, another conceivable explanation for these fatty acids with a comparable immature CPI_{FA} signature at this advanced level of maturity is the incorporation of bacterial biomass from deep microbial communities containing C_{16} and C_{18} fatty acids as main cell membrane components. Microbial life might have been stimulated at this interval by the increasing thermally-induced release of potential substrates from the organic matrix during early catagenesis.

The lignite and coal matrices were also analysed by open system pyrolysis-GC before and after saponification. The pyrolysis chromatograms of the low rank samples show that the kerogen-linked alcohols are detectable as their corresponding alkenes in open system pyrolysis.

6.6. Acknowledgements

We are grateful to Cornelia Karger and Ferdinand Perssen (GFZ Potsdam, Germany) for GC-MS and pyrolysis-GC measurement and Richard Sykes (GNS Science, New Zealand) for providing basic information on the sample material. We thank N. Gupta and P. Schaeffer for reviewing this paper and for providing helpful suggestions to improve the manuscript. Finally, we would like to thank the Helmholtz Association for the financial support of this study.

6.7. Additional section: Additional compounds identified in the neutral polar fraction

Hydroxy-fatty acids

ω-hydroxy fatty acids are related to the plant waxes suberin and cutin deriving from protective coatings from plant tissues. They form cross-linked structures by the formation of ester bonds (Blokker et al., 1998, 2000) and have mainly chain lengths between 16 and 28 carbon atoms.

ω-hydroxy fatty acids were only found in the very immature samples of the Northland Basin (G001986 and G001988) and the Eastern Southland Basin samples (G001976 and G001978) (Figure 6.7, Table 6.2). Only hydroxy fatty acids with an even carbon number chain length in the range of 16 to 28 carbon atoms were detected. The C_{16} acid was found to be the most abundant acid. In the longer carbon chain length acids (C_{18}-C_{28}), a second maximum at C_{24} for the Northland Basin samples and at C_{26} for the Eastern Southland Basin samples can be observed. In the late diagenesis and catagenesis stage samples, ω-hydroxy fatty acids could not be detected anymore most likely due to a rapid degradation during early maturation.

6.7. Additional section: Additional compounds identified in the neutral polar fraction

Dicarboxylic acids

It was reported that α-ω-dicarboxylic acids were found in young and shallow subsurface peat samples in the upper first cm but these compounds almost dissappear completely in deeper and older deposits as a result of rapid degradation (Disnar et al., 2005).

α-ω-dicarboxylic acids were only detected in the two Northland Basins samples (G001986 and G001988) showing an even over odd carbon number predominance (Figure 6.8, Table 6.2). Therefore, their occurrence appear to be highly facies dependent. In these samples, the higher plant deriving acids with chain length of 18 to 28 carbon atoms as well as a C_9 acid were detected. Interestingly, Grasset et al. (2002) investigating the acid distribution in a calcidic peat sample (Marais Poitevin, Deux-Sèvres, Western part of France) were able to detect dicarboxylic acids with the same carbon number distribution including the C_9 acid. In sample G001986, the dicarboxy-FAs

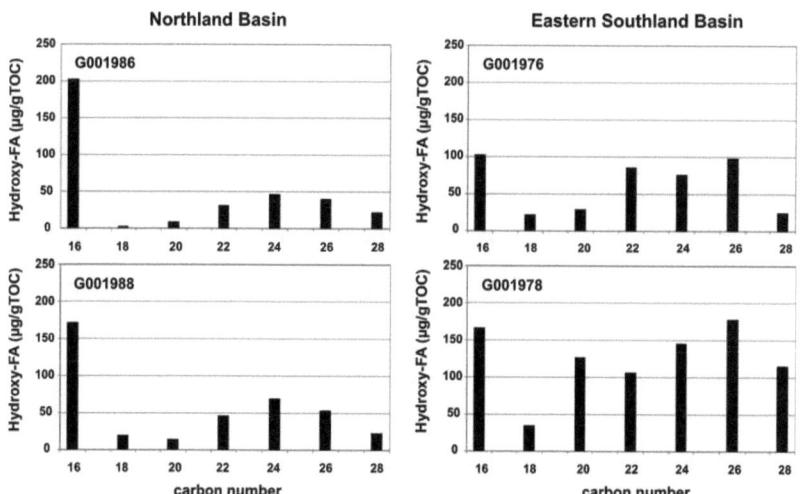

Figure 6.7.: *Distribution of ω-hydroxy fatty acids detected in the Northland and Eastern Southland Basin samples after saponification. Numbers indicate the chain length of the ω-hydroxy-FA.*

6. Ester-Bound Fatty Acids And Alcohols Linked To The Kerogen

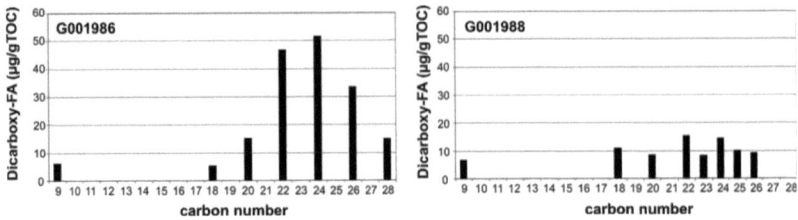

Figure 6.8.: *Distribution of α-ω-dicarboxylic acids found in the Northland Basin samples after saponification. Numbers indicate the cahin lenght of the dicarboxy-FA.*

are up to 4 times higher than in sample G001988. Except the C_9 acid, no odd carbon numbered dicaboxy-FAs were detected in this sample. However, in sample G001988, odd carbon numbered dicarboxy-FAs appeared. The most abundant α-ω-dicarboxylic acid is the C_{24} (51.61 μg/gTOC, G001986, Table 6.2). Incorporated into the organic matrix, the dicarboxylic fatty acids most likely play a role in the formation of link-bridges esterified with e.g. OH-groups of phenolic compounds (Blokker et al., 2001; Petersen et al., 2008).

The sources of long chain (C_{16+}) dicarboxy acids are also most probably cutin and suberin from higher plant cuticles (Kolattukudy et al., 1976) or microbial oxidation of fatty acids via ω-hydroxy carboxylic acids. The C_9 dicarboxy acid probably results from the oxidation of the carbon-carbon double bonds of Δ-9 fatty acids (Amblès et al., 1994; Grasset et al., 2002).

Functionalised pentacyclic triterpenoids

Functionalised pentacyclic triterpenoids bound to the kerogen were only found in the immature Eastern Southland samples. Betulin (Lup-20(29)-en-3β,28-diol) was found in high amounts in both samples (G001976: 2.38 mg/gTOC; G001978: 1.11 mg/gTOC, Table 6.2). In samples G001978 also lupeol (Lup-20(29)-en-3β-ol) could be detected (0.16 mg/gTOC, Table 6.2) (Figure 6.9). Betulin is found in the bark of birch trees (*Betulaceae*). High amounts of betulin are usually found in *Betulaceae*-dominated peats (Köller,

6.7. Additional section: Additional compounds identified in the neutral polar fraction

	G001986	G001988	G001976	G001978
Dicarboxylic acids [µg/gTOC]				
n-C_9	6.23	6.85	-	-
n-C_{18}	5.2	11.1	-	-
n-C_{19}	-	-	-	-
n-C_{20}	15.3	8.66	-	-
n-C_{21}	-	-	-	-
n-C_{22}	46.79	15.47	-	-
n-C_{23}	-	8.33	-	-
n-C_{24}	51.61	14.58	-	-
n-C_{25}	-	10.17	-	-
n-C_{26}	33.45	9.26	-	-
n-C_{27}	-	-	-	-
n-C_{28}	14.95	-	-	-
ω-hydroxy acids [µg/gTOC]				
n-C_{18}	2.42	19.31	22.15	34.4
n-C_{20}	8.78	14.60	29.44	127.12
n-C_{22}	31.36	46.41	86.11	106.03
n-C_{24}	46.24	68.94	76.52	145.68
n-C_{26}	39.81	53.47	98.1	178.47
n-C_{28}	21.98	23.04	24.43	115.2
Functionalised pentacyclic triterpenoids [µg/gTOC]				
Betulin	-	-	2380	1110
Lupeol	-	-	-	160

Table 6.2.: *Concentrations of dicarboxylic acids, ω-hydroxy acids and functionalised pentacyclic triterpenoids detected in the NZ lignite samples.*

6. Ester-Bound Fatty Acids And Alcohols Linked To The Kerogen

2002). However, the birch tree is not endemic in New Zealand. Another possible betulin source may be the endemic ribbonwood (*Plagianthus betulinus*, also called New Zealand cotton) (Sakai and Wardle, 1978; Devine, 1982).

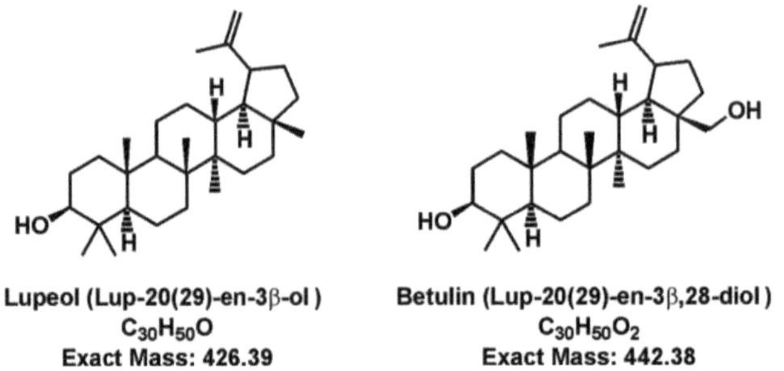

Lupeol (Lup-20(29)-en-3β-ol)
$C_{30}H_{50}O$
Exact Mass: 426.39

Betulin (Lup-20(29)-en-3β,28-diol)
$C_{30}H_{50}O_2$
Exact Mass: 442.38

Figure 6.9.: *Structures of the pentacyclic triterpenoid alcohols betulin and lupeol.*

Conclusion

Fatty acids bearing additional functional groups like the detected α-ω-dicarboxylic acids and the ω-hydroxy carboxylic acids are able to form cross-linked structures within the kerogen by the formation of ester and ether bonds with hydroxy groups of the penolic structurs from sinapyl-, coumaryl- and conifery alcohol. The data indicate that these structures were rapidly degradated during early diagenesis because they were only found in the immature lignin samples from the Northland and the Eastern Southland Basins.

The high amouts of betulin found in the Eastern Southland Basin samples may point to a high contribution of ribbonwood to the Eastern Southland lignites.

7. Free Fatty Acids, Alcohols And Esters From The Bitumen Fraction

7.1. Introduction

The bitumen fraction of a sediment sample is defined as the organic matter soluble in common organic solvents (Durand, 1980). Therefore, bitumen consists of free, unbound organic substances mainly deriving from former plant material or the remains of microorganisms. It represents a highly complex mixture of different organic compounds containing aromatic and aliphatic constituents as well as compounds containing heteroatoms such as free carboxylic acids, alcohols or esters.

The bitumen can be fractionated into several compound fractions according to their polarity or chemical behavior. Bitumen or oil show a good solubility in polar organic solvents e.g. in dichloromethane which is, therefore, used as a standard solvent for extraction. By addition of an excess of light hydrocarbons (pentane, hexane or heptane), the so-called asphaltenes (molecules with a higher complexity) precipitate from the bitumen (Mitchell and Speight, 1973; Speight et al., 1984; Leontaris and Mansoori, 1988; Theuerkorn et al., 2008) and the compounds remaining in solution are named maltenes. Maltenes can further be separated by column chromatography into fractions of different polarity such as the aliphatic, aromatic and

7. Free Fatty Acids, Alcohols And Esters From The Bitumen Fraction

the NSO fraction (nitrogen, sulfur and oxygen containing polar compounds). Asphaltenes are thought to show a similar characteristic than the parent kerogen itself. Therefore, it is discussed in the literature that petroleum asphaltenes can provide insights into the constitution of unexplored or even unreachable deeply buried source rocks (Horsfield et al., 1991; Di Primio et al., 2000; Keym et al., 2006; Lehne and Dieckmann, 2007).

Both the kerogen and bitumen of a source rock may contribute to petroleum formation, especially in carbonate and phosphatic source rocks where the bitumen content is high (typically 120 mg/gTOC) (Palacas et al., 1984; Powell, 1984; di Primio and Horsfield, 1996). In humic coals also the bitumen content is high (Allan et al., 1977). Values of 90mg/gTOC (Horsfield, 1984; Hvoslef et al., 1988; Bechtel et al., 2005) and even up to 200mg/gTOC (Sykes and Snowdon, 2002) have been reported. The aliphatic and aromatic hydrocarbon fractions of the bitumen is well characterised in order to understand the petroleum formation reactions (Schenk et al., 1997). Levine (1993) suggested that low rank bitumen might progressively become insoluble by reactions with macromolecular coal components. It was suggested by Mansuy et al. (1995) that bitumen may act as hydrogen transfer agent and that its presence or absence is influencing the degree of condensation and cross linking reactions. Also for the New Zealand coals, the role of bitumen affecting the composition of high temperature pyrolysis products by stabilisation of free radicals and prevention of recombination processes was demonstrated (Vu et al., 2008).

Oxygen functionalised compounds from the bitumen were also characterised in detail especially in the focus of fossil fuels describing the occurrence of alkylphenols (Bennett et al., 1996; Simoneit et al., 1996), carboxylic acids (Seifert, 1975; Barakat and Rullkötter, 1995), chromanes (Sinninghe Damsté et al., 1993), ketones and aldehydes (Wang and Rullkötter, 1997) in relation to maturity. Cranwell (1977) e.g. documented the occurrence of 2-alkanones with significant odd-over-even predominace attributed to the biological transformation of fatty acids by β-oxidation and subsequent

7.1. Introduction

decarboxylation. Wang and Rullkötter (1997) described aliphatic aldehydes as constituents of the Green River Formation oil shale and long chain aliphatic aldehydes have also been identified in marine sediments by Prahl and Pinto (1987). The occurrence and distribution of several types of aromatic ketones and aldehydes analysed from the NSO fraction of bitumen from the Posidonia Shale formation is described in detail in Wilkes et al. (1998). In this study, they introduced the 1-indanone/(1-indanone + 1-tetralone) ratio (ITR) as a novel heterocompound based maturity parameter.

Furthermore, the relation of bitumen constituents and kerogen-bound components has been the subject of several investigations. Similarities in the distribution of fatty acids from bitumen and kerogen-bound fractions of sulfur-rich sediments from the Nördlinger Ries (Germany) were reported e.g. in Barakat and Rullkötter (1995). However, Farrington and Quinn (1971) showed variations in the fatty acid constitution of short sub-surface cores from recent sediments from Narragansett Bay (Rhode Island, USA) and speculated on differences in the susceptibility to degradation of bitumen and kerogen-bound compounds.

In this chapter, the composition of free fatty acids, alcohols and esters in the bitumen obtained from New Zealand coal samples of different maturity (R_0: 0.28% to 0.8%) are investigated and subsequently, the results are compared with the kerogen bound fatty acids and alcohols obtained after alkaline hydrolysis of the pre-extracted coal samples (*cf.* Chapter 6). This comparison adresses differences and similarities in the constitution of bitumen and kerogen and is used to discuss the sources of these compounds, their incorporation into the kerogen matrix and the relation of kerogen and bitumen within coals and lignites during diagenesis and early catagenesis.

7. *Free Fatty Acids, Alcohols And Esters From The Bitumen Fraction*

7.2. Materials and methods

7.2.1. Sample material

For this study, 10 coal samples from different coal mines in New Zealand and two lignite samples from the DEBITS-1 well were selected. The coal mine samples include samples G001976 and G001978 from the Eastern Southland Basin, G001980, G001982 and G001983 from the Waikato Basin, G001989, G001993, G001995 and G001996 from the West Coast Basin and the sample G001991 from the Taranaki Basin. Information about coalfields, formation, age and thermal maturity are listed in table 3.1. Additional geochemical data for these samples can be found in table 3.2.

The two additional lignite samples taken from the DEBITS-1 well are G004541 (18.86-19.02 m) and G004544 (63.41-63.51 m), both from the Tauranga Group. Further information about these samples can be found in table 3.3.

7.2.2. Sample preparation

The coal samples (ca. 9 g) were extracted with 200 ml of an azeotropic mixture of acetone (38 %), chloroform (32 %) and methanol (30 %) using a Soxhlet device for 3 d. The solvent was reduced to a volume of 100 ml and 10 ml of the obtained dissolved bitumen fraction was separated for further treatement.

To this aliquot, 10 μg of both, erucic acid and androstanone were added as internal standards for subsequent quantification. The solvent was removed and the bitumen was again dissolved in 1 ml of the azeotropic mixture using ultrasonication. For precipitation of the asphaltenes, 40 ml of n-hexane was added to this solution and the mixture was placed for 15 min in an ultrasonic bath. The mixture was than filtered using a funnel equipped with pre-extracted cotton wool and Na_2SO_4 to separate dissolved

7.2. Materials and methods

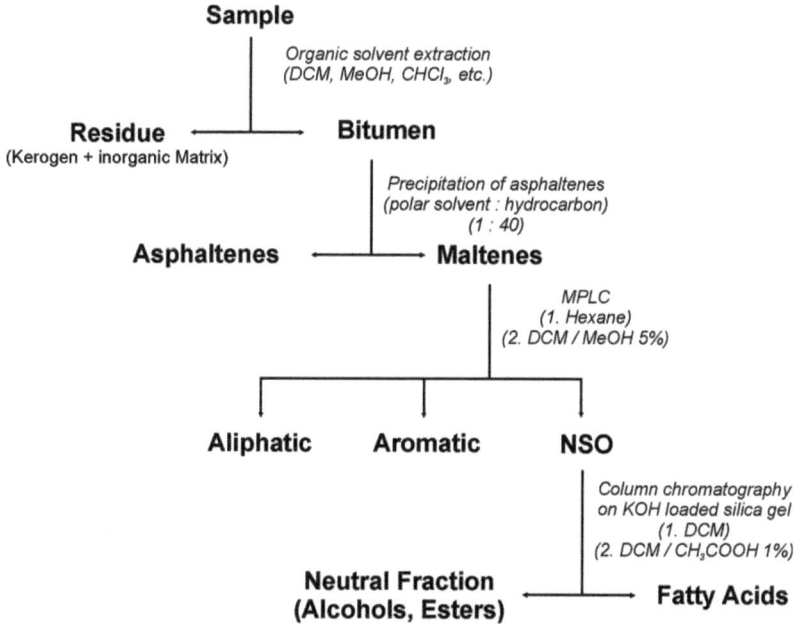

Figure 7.1.: *Stepwise procedure for the fractionation of bitumen for the analysis of free fatty acids, alcohols and esters. MPLC: medium pressure liquid chromatography (Radke et al., 1980), NSO: nitrogen, sulfur and oxygen fraction.*

maltenes and precipitated asphaltenes. The solvent n-hexane was than removed from the maltene fraction at room temperature using a TurboVap® device (Zymark, divison of Caliper Life Sciences GmbH, Germany) and finally a nitrogen gas stream.

The fractionation of the maltenes was achieved using a MPLC (medium pressure liquid chromatography) system from MKW Chromatography (Jülich, Germany) according to the procedure described in Radke et al. (1980) and Willsch et al. (1997). For this purpose the samples were dissolved in 1 ml n-hexane and injected. For the elution of the aliphatic and aromatic compounds n-hexane was used, the NSO fraction was eluted with a mixture of dichloromethane containing 5% methanol.

7. Free Fatty Acids, Alcohols And Esters From The Bitumen Fraction

Subsequently, the NSO fraction was separated by column chromatography into a fatty acid and a neutral polar fraction according to the method described in chapter 4.1.4 for the separation of the saponification products. For the chromatography, a KOH impregnated silica gel column was used as stationary phase. The neutral fraction (containing alcohols and esters) was eluted with DCM and the fatty acids were eluted with a mixture of DCM and 1% formic acid. After removal of the solvent, the neutral fraction was treated with N-Methyl-N-(trimethylsilyl)fluoroacetamide (MSTFA) and the fatty acid fraction with diazomethane for derivatisation prior to GC-MS analysis. A scheme for the separation procedure is shown in figure 7.1.

7.3. Results and discussion

7.3.1. Fatty acid fraction

The free fatty acids separated from the bitumen show a bimodal distribution with a strong even over odd carbon number predominance. The first maximum is located at palmitic acid (16 carbon atoms) and the second maximum is located between chain lengths of 24 and 26 carbon atoms. Figure 7.2 shows an example of a typical fatty acid distribution in an immature coal sample from the Eastern Southland Basin (G001976).

The high amounts of long chain fatty acids with even preference are characteristic for higher terrestrial plant supply (Eglinton and Hamilton, 1967) while the n-C_{16} and n-C_{18} fatty acids might have many different sources since they are common not only in bacteria (Ratledge and Wilkinson, 1988), but also constituents in eukaryotes (Cranwell, 1974; Shameel, 1990; Oldenburg et al., 2000).

The total amounts of free fatty acids show a strong decrease from 14 mg/gTOC (sample G004544) down to approximately 0.2 mg/gTOC (samples G001991 and G001993) with ongoing maturation. The main decrease was

7.3. Results and discussion

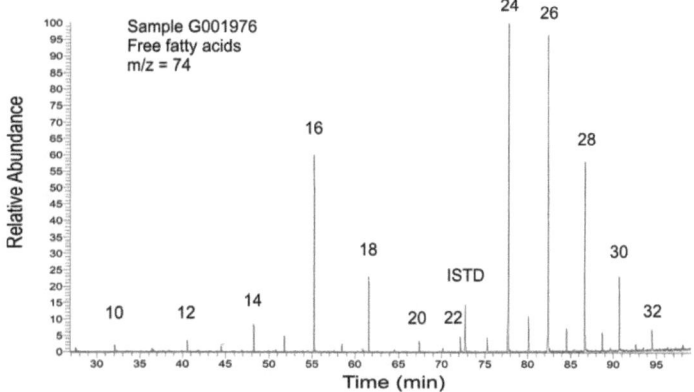

Figure 7.2.: *Distribution of free fatty acids in the bitumen of the immature sample G001976 from the Eastern Southland Basin (m/z=74). Numbers indicate carbon number chain length of fatty acids. ISTD: internal standard (erucic acid)*

observed during diagenesis and early catagenesis with exception of the two Eastern Southland Basin samples (G001976 and G001978) which show significant smaller amounts (Figure 7.3). A reason for this might be facies related differences in the initial plant material.

Separation of the fatty acid concentrations into the short chain fatty acids (n-C_{14} to n-C_{19}) and terrestrial plant derived long chain fatty acids (n-C_{20} to n-C_{32}) revealed that the main part of the fatty acid profile (Figure 7.3) in the bitumen is caused by the signal of the terrestrial derived long chain fatty acids. Their amounts are up to 4 times higher than those of the short chain fatty acids. Therefore, the short chain fatty acids are of minor importance within the bitumen fraction, although their abundance versus maturity shows a similar trend compared to the long chain fatty acids.

Regarding the short chain fatty acids, the two Eastern Southland samples show comparatively high values. In contrast, the terrestrial long chain fatty acids of these samples have unusual low amounts actually being responsible for the low amounts observed for the total fatty acids (Figure 7.3).

7. Free Fatty Acids, Alcohols And Esters From The Bitumen Fraction

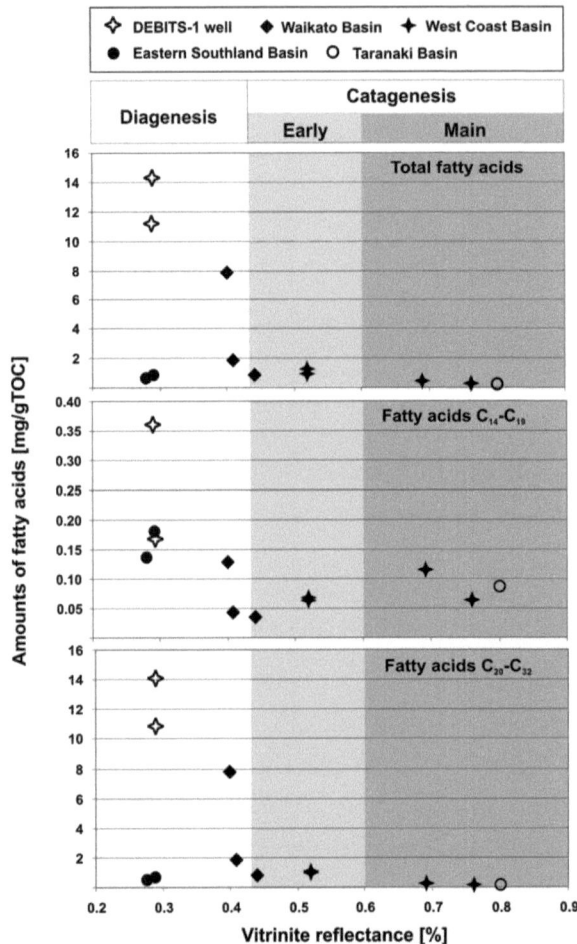

Figure 7.3.: *Concentration trend of the total free fatty acids in the bitumen fraction of the NZ coal samples of different maturity and additionally in total and separated into short chain (n-C_{14} to n-C_{19}) and long chain fatty acids (n-C_{20} to n-C_{32}).*

This observation seems to support the suggestion that facies variation may have a high influence on the total amounts of the fatty acids within the coal samples especially in the lower maturity range. In the higher maturity

7.3. Results and discussion

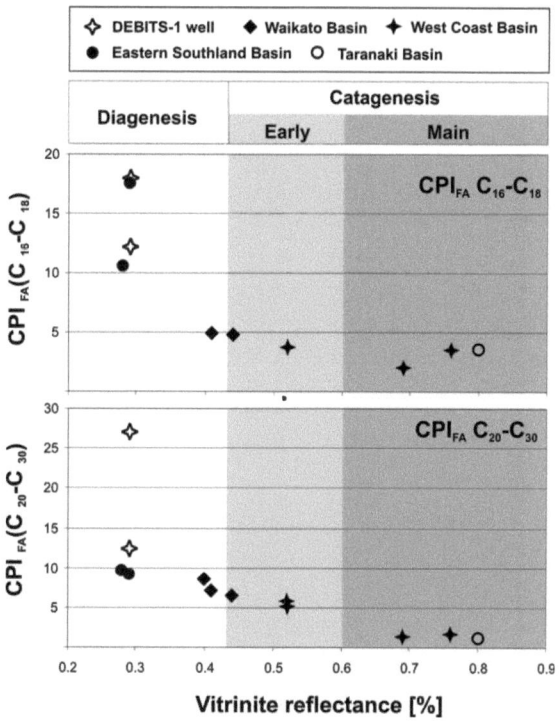

Figure 7.4.: *Calculated carbon preference index for fatty acids (CPI_{FA}) separated for short chain and long chain fatty acids (FAs) vs. maturity.*

range, variations in the samples become smaller due to the advanced level of organic matter degradation (*cf.* Chapter 6).

As described for the kerogen bound fatty acids and alcohols in chapter 6, a CPI_{FA} (carbon preference index for fatty acids) was calculated for the short chain and long chain fatty acids separately, to obtain a maturation related signal that is less influenced by facies related differences (Figure 7.4). In sample G001982 from the Waikato Basin, the concentrations of the odd fatty acids n-C_{15}, n-C_{17} and n-C_{19} were below detection limit and, therefore, no CPI_{FA} could be calculated for the short chain fatty acids of this

7. Free Fatty Acids, Alcohols And Esters From The Bitumen Fraction

Sample	Basin/Well	R_0	Total FAs	FA C_{14}-C_{19}	FA C_{20}-C_{32}	CPI_{FA} C_{16}-C_{18}	CPI_{FA} C_{20}-C_{30}
G001978	Eastern Southland	0.28	0.71	0.14	0.57	10.6	9.6
G001976	Eastern Southland	0.29	0.84	0.18	0.66	17.6	9.1
G004544	DEBITS-1	0.29	14.25	0.17	14.08	12.2	12.6
G004541	DEBITS-1	0.29	11.18	0.36	10.83	17.9	27.2
G001982	Waikato	0.40	7.82	0.13	7.69	n.d.	8.5
G001983	Waikato	0.41	1.87	0.04	1.83	4.9	7.2
G001980	Waikato	0.44	0.83	0.03	0.8	4.8	6.5
G001995	West Coast	0.52	1.05	0.06	0.98	3.7	5.1
G001996	West Coast	0.52	1.14	0.07	1.07	3.7	5.8
G001989	West Coast	0.69	0.39	0.12	0.27	2.1	1.3
G001993	West Coast	0.76	0.22	0.06	0.15	3.5	1.7
G001991	Taranaki	0.80	0.24	0.09	0.15	3.6	1.2

Table 7.1.: *Total amounts of free fatty acids from the bitumen fraction (in mg/gTOC) and calculated CPI_{FA} values separated for short chain and long chain fatty acids. n.d.: not determined (odd FA are below detection limit)*

sample.

The calculated CPI_{FA} values for both the short and long chain length fatty acids show a decrease with ongoing maturation, while the main decrease is observed during diagenesis and early catagenesis. Although the total amounts of long chain fatty acids in the two Eastern Southland samples are relatively low, their CPI_{FA} values are comparatively high (9.1-9.6), representing the expected immature material plotting into the general trend outlined by the other samples and confirming that differences in the facies are an explanation for the low fatty acid amounts in these samples. With increasing maturity, the long chain fatty acids show CPI_{FA} values decreasing close to unity in the main catagenesis samples from the Taranaki and West Coast Basins. The CPI_{FA} values for the short chain fatty acids initially also decrease, but remain constant between 2 and 4 during the main catagenesis stage. The calculated values for the sample set are listed in table 7.1.

7.3.2. Neutral fraction (alcohols, esters and functionalised pentacyclic triterpenoids)

In the neutral polar fraction free, alcohols as well as fatty acid methyl esters (FAME) and pentacyclic triterpenes were detected. In general, the amounts of compounds in the neutral fraction are 1-2 orders of magnitude lower than those of the free fatty acids. With respect to the fact that no methylation was applied to these samples, the FAME must have been a part of the bitumen.

Sample	Basin/Well	R_0	ALC (total)	FAME (total)	FAME C_{14}-C_{19}	FAME C_{20}-C_{32}	CPI_{FAME} C_{16}-C_{18}	CPI_{FAME} C_{20}-C_{30}
G001978	E. Southland	0.28	23.7	2.0	0	2.0	n.d.	8.3
G001976	E. Southland	0.29	94.4	67.1	6.9	60.2	13.8	8.4
G004544	DEBITS-1	0.29	655.4	127.9	1.0	126.9	3.7	9.6
G004541	DEBITS-1	0.29	339.8	30.7	0.7	30	n.d.	17.8
G001982	Waikato	0.40	1.2	84.3	3.5	80.7	2.1	5.7
G001983	Waikato	0.41	0.5	19.0	1.0	18.0	1.9	5.8
G001980	Waikato	0.44	0.2	8.9	0.5	8.4	2.3	5.1
G001995	West Coast	0.52	0.2	7.3	1.2	6.1	1.7	3.3
G001996	West Coast	0.52	0.3	6.3	0.8	5.4	2.1	3.7
G001989	West Coast	0.69	0	0	0	0	n.d.	n.d.
G001993	West Coast	0.76	0	3.1	0.7	2.4	1.6	1.2
G001991	Taranaki	0.80	0	0	0	0	n.d.	n.d.

Table 7.2.: *Total amounts of free alcohols (ALC) and fatty acid methyl esters (FAME) from the bitumen fraction (in µg/gTOC) and calculated CPI_{FAME} values separated for short chain and long chain fatty acid methyl esters. n.d.: not determined (FAME below detection limit)*

Table 7.2 shows the amounts of alcohols and FAME obtained from the bitumen and the calculated CPI_{FAME} (Carbon Preference Index for fatty acid methyl esters). In the relatively mature samples G001991 (Taranaki Basin) and G001989 (West Coast Basin), neither alcohols nor FAME could be detected. In some cases (samples G004541 and G001978), the concentrations of the n-C_{15}, n-C_{16} and n-C_{19} compounds were below detection limit. There-

7. Free Fatty Acids, Alcohols And Esters From The Bitumen Fraction

fore, no CPI_{FAME} values were determined for these samples.

Alcohols

Significant amounts of free alcohols were only detected in the very immature samples from the DEBITS-1 well and from the Eastern Southland Basin being in the maturity range of early diagenesis. In the early and especially in the main catagenesis stage samples almost no alcohols could be detected (Table 7.2, Figure 7.5).

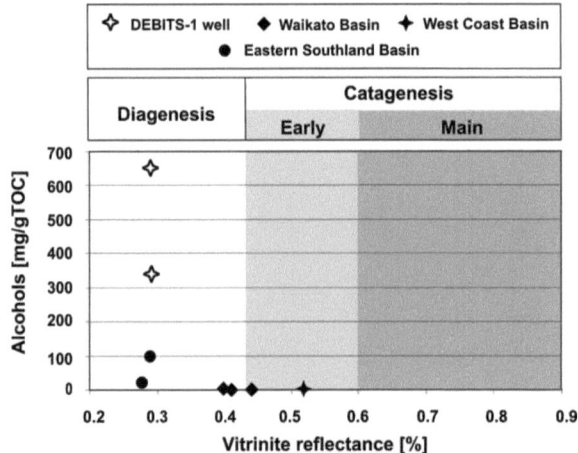

Figure 7.5.: *Concentrations of the total alcohols in the bitumen fraction of the NZ coal samples of different maturity.*

The distribution pattern of the alcohols shows a strong even over odd carbon number predominance due to their biosynthesis from the corresponding fatty acids by reduction. Alcohols obtained from the DEBITS-1 well samples have concentrations which are about one order of magnitude higher than those obtained from the Eastern Southland samples, most likely representing different facies. The distribution pattern of these alcohols (Figure 7.6) reveal alcohols ranging between 22 to 30 carbon atoms

7.3. Results and discussion

with a strong even carbon number predominance and a maximum at the n-C_{24} alcohol. The obtained distribution is typical for terrestrial derived plant material (Eglinton and Hamilton, 1967).

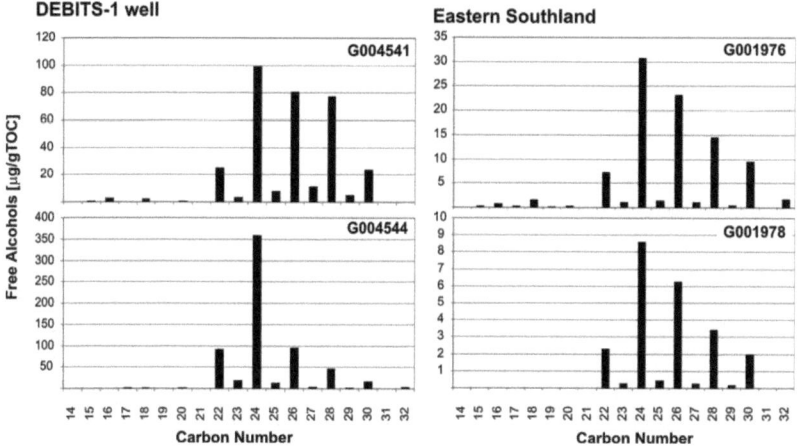

Figure 7.6.: *Distribution of the free alcohols in the bitumen of the immature samples from the DEBITS-1 well (Waikato Basin) and from the Eastern Southland Basin.*

Fatty acid methyl esters

Fatty acid methyl esters (FAME) were found to be part of the bitumen fraction in almost all samples investigated, with exception of two samples from the Taranaki Basin (G001991) and the West Coast Basin (G001989) (Table: 7.2). The distribution pattern of the FAMEs shows high concentrations in the long chain length range with an even over odd carbon number predominance and a maximum at n-C_{26}, representing terrestrial plant supply. Separation of the short chain FAMEs (n-C_{14} to n-C_{19}) and the long chain FAMEs (n-C_{20} to n-C_{32}) revealed that the concentrations, as already observed for the free fatty acids, are mainly dominated by the long FAMEs (Figure 7.7).

The amounts of FAMEs decrease with ongoing maturation. Main decrease is also observed during diagenesis. It can be seen in figure 7.7 that

7. Free Fatty Acids, Alcohols And Esters From The Bitumen Fraction

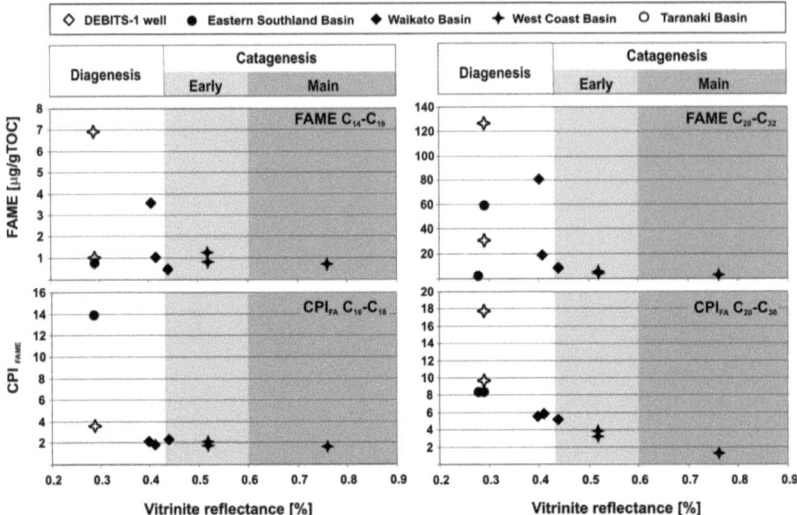

Figure 7.7.: *Distribution of the fatty acid methyl esters (FAME) and the calculated CPI_{FA} in the bitumen of NZ coal samples, separated for long chain and short chain esters.*

the immature samples in the stage of diagenesis show also (as for the free FAs) a strong variability in the short and long chain length range of the FAMEs which again might be related to facies variations. In the calculated CPI_{FAME}, expected to be less facies dependent (Glombitza et al., 2009a), a clear maturation related decrease is revealed, reflecting the ongoing decrease of the esters during diagenesis and early catagenesis. The sample of highest maturity containing FAMEs was sample G001993 from the West Coast Basin showing a CPI_{FAME} value closed to unity for the long chain FAMEs.

FAMEs in the bitumen fraction may either derive from methylation of free fatty acids within the sediment or they may derive from metylation of free fatty acids during the soxhlet extraction with the azeotrope mixture that contains 30% methanol. The extract of the coal samples are slightly acidic and esterification may appear during the three day heating procedure in the soxhlet device. This could be the reason why no other alkyl fatty acid esters

7.3. Results and discussion

than methyl ester could be detected in the bitumen fraction. Furthermore, the amouts of the FAMEs are significantly lower than those of the free fatty acids. This also might point to the suggestion that the free fatty acids may be precursors of the detected FAMEs. Therefore, the same origin for the FAMEs as for the free fatty acids is assumed.

Functionalised pentacyclic triterpenoids

Functionalizsed pentacyclic triterpenoids were detected in the samples from the Eastern Southland Basin (G001976 and G001978) being in the diagenetic stage and in three samples from the Waikato Basin (G001980, G001982 and G001983) being in the stage of late diagenesis to early catagenesis (Table 7.3). Beyond the early catagenesis, no functionalised pentacyclic triterpenoids were detected in the NZ coal samples anymore.

Compound	G001978 (R_0: 0.28%)	G001976 (R_0: 0.29%)	G001982 (R_0: 0.40%)	G001983 (R_0: 0.41%)	G001980 (R_0: 0.44%)
Taraxerol	-	33.7	-	-	-
Taraxerenone	-	28.4	-	-	-
β-Amyrin	5.7	86.2	-	-	-
Oleanenone/Ursenone	20.2	51.0	-	-	-
δ-Amyrin	5.2	18.5	-	-	-
Lupenone	4.7	26.1	-	-	-
Friedelin	8.5	20.0	-	-	-
U1	-	-	62.0	9.3	7.2
U2	2.6	9.3	-	-	-
U3	5.7	5.3	42.8	20.6	4.7

Table 7.3.: *Pentacyclic triterpenoids detected in the Eastern Southland Basin (G001978 and G001976) and three Waikato Basin samples (G001982, G001983, G001980) in $\mu g/gTOC$. U1, U2, U3 = unknown terpenoid.*

Pentacyclic triterpenoids can be used as indicators for the initial plant contributors (Ives and O'Neill, 1958; Baas et al., 2000; Köller, 2002). They are constituents of resins and cuticular waxes in higher terrestrial plants (Gülz, 1994; Köller, 2002). The distribution of thiese pentacyclic triterpenoids in the NZ coal samples appears to be strongly related to the different basins and sees to reflect different vegetation sources. In the Eastern Southland Basin samples, five different pentacyclic triterpenoids (β-amyrin, oleanenone/ursenone, δ-amyrin, lupenone and friedelin, Figure 7.9) could be identified. Taraxerol and taraxerenone (Figure 7.9) were only detected in sample G001976. Two unidentified terpenoids (U2 and U3) of relatively high abundance were also found in these samples.

The Waikato Basin samples only contain two unidentified terpenoids of similar high abundance. One is compound U2, also found in the Eastern Southland Basin samples and the second terpenoid is U1, only detected in the Waikato Basin samples. Figure 7.8 shows the mass spectra of the three unknown terpenoids. Due to the fragments m/z = 203 and 424 the terpenoid U1 is a ketone with an unsaturation within the C/D ring system (Figure 7.9).

Taraxerol, taraxerone and β-amyrin are often found in leaves of plants from mangrove swamps of the *Rhizophoraceae* species (W. Hogg and T. Gillan, 1984; Killops and Frewin, 1994). Friedelin is not restricted to any particular plant family and was found in lichens and in peat as well as brown coal and is often accompanied by β-Amyrin (Sainsbury, 1970). In general, the terpenoids of leaf cuticules occur together in different combinations. For the identification of the distributing plant families and species, a more detailed investigation of the terpenoids would be necessary. Statistical cluster analysis would be an appropriate tool to receive a deeper insight into the contributing plant material to classify the investigated lignites and coals into groups (W. Hogg and T. Gillan, 1984; Köller, 2002).

7.3. Results and discussion

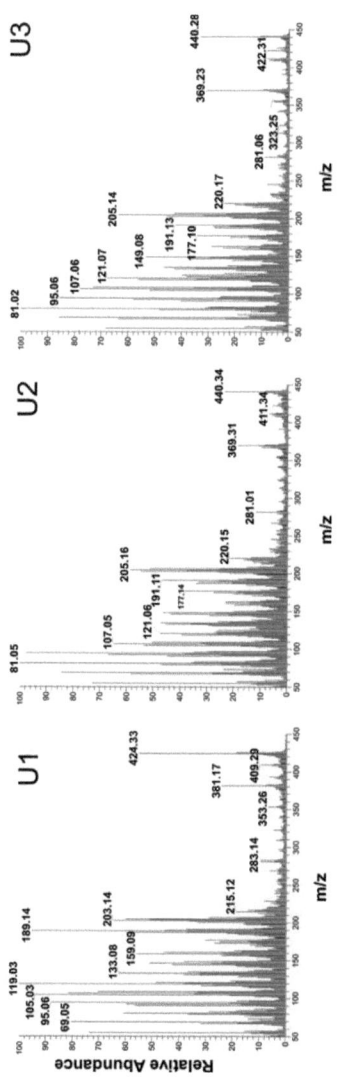

Figure 7.8.: *Mass spectra of the unknown pentacyclic triterpenoids U1, U2 and U3 detected in the neutral fraction of the bitumen extract from the NZ coal samples.*

7. Free Fatty Acids, Alcohols And Esters From The Bitumen Fraction

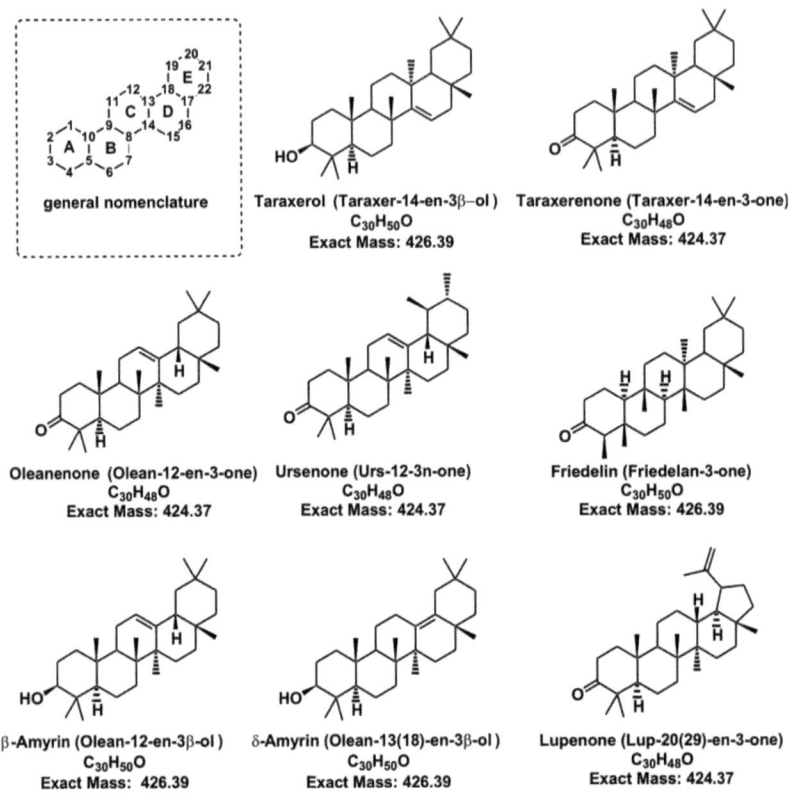

Figure 7.9.: *Structures of identified pentacyclic triterpenoids Taraxerol, Taraxerenone, Oleanenone, Ursenone, Friedelin, β-Amyrin, δ-Amyrin and Lupenone.*

7.3.3. Comparison of free and kerogen bound n-fatty acids, alcohols and functionalised terpenoids

n-Fatty acids

The total amounts of fatty acids extracted from the bitumen of the NZ coal samples are more or less in the same range as the kerogen bound fatty acids usually a little higher (cf. Tables 6.1 and 7.1). However, the DEBITS-1 well samples show very high amounts of free fatty acids (cf. Figure 7.3, Table 7.1) which are up to 7 times higher than the amounts of the corresponding kerogen bound fatty acids (cf. Figure 8.2, Table 8.2). Furthermore, it has to be considered that some FAMEs were detected in the bitumen fraction (Figure 7.7, Table 7.2) which presumably have to be added to the free fatty acids additionally enlarging the bitumen fatty acid fraction.

Sample	Basin/Well	R_0 [%]	Bound FA fraction		Free FA fraction	
			SCFA [%]	LCFA [%]	SCFA [%]	LCFA [%]
G001978	E. Southland	0.28	72.4	21.6	19.7	80.3
G001976	E. Southland	0.29	38.8	46.9	21.4	78.6
G004544	DEBITS-1	0.29	36.7	50.2	1.2	98.8
G004541	DEBITS-1	0.29	32.3	63.7	3.2	96.9
G001982	Waikato	0.40	59.5	34.4	1.7	98.3
G001983	Waikato	0.41	36.7	48.8	2.1	97.9
G001980	Waikato	0.44	28.9	68.5	3.6	96.4
G001995	West Coast	0.52	75.0	23.1	5.7	93.3
G001996	West Coast	0.52	78.4	20.0	6.1	93.9
G001989	West Coast	0.69	74.7	19.6	30.8	69.2
G001993	West Coast	0.76	49.8	37.4	27.3	68.2
G001991	Taranaki	0.80	57.7	29.9	37.5	62.5
Mean Value			53.4	38.7	13.4	86.2

Table 7.4.: *Percentage proportions of short chain fatty acids (SCFA, $C_{12/13}$ to C_{19}) and long chain fatty acids (LCFA, C_{20} to $C_{30/32}$) of the amount of total fatty acids in the kerogen-bound and the free (bitumen) fatty acid fraction. (Note: Difference to 100% in the kerogen-bound fraction is due to the occasional occurrence of n-C_6 to n-C_{12} fatty acids which are not considered here or in the parameters of the previous chapters.)*

The distribution pattern of the free and the kerogen bound fatty acids is to some extent similar (cf. Figures 6.2 and 7.2), with a bimodal distribution showing maxima at n-C_{16} and n-C_{24} or n-C_{26}. However, the amounts of the short chain fatty acids, mainly n-C_{16} and n-C_{18}, are somewhat lower in the free than in the kerogen-bound fatty acid fraction (cf. Figures 6.5 and 7.3;

7. Free Fatty Acids, Alcohols And Esters From The Bitumen Fraction

Tables 6.1 and 7.1). In contrast, the plant wax derived long chain fatty acids (n-C_{20} to n-$C_{30/32}$) are significantly higher in the bitumen than in the bound fraction (*cf.* Figures 6.5 and 7.3; Tables 6.1 and 7.1). As a result, the relative proportions of short chain to long chain fatty acids differ between the free and the bound fraction. While the short chain fatty acids dominate the fatty acid distribution in the bound fraction, the long chain fatty acids are the dominating compounds of the free fatty acids (Table 7.4). For example, the early catagenesis sample G001996 (R_o 0.52%, West Coast Basin) contains 1.05 mg/gTOC bound fatty acids (Table 6.1) and with 1.14 mg/gTOC almost the same amount of free fatty acids (Table 7.1) but their relative proportions within the fatty acid distribution is significantly different. With 0.21 mg/gTOC, the plant wax deriving long chain fatty acids (n-C_{20} to n-$C_{30/32}$) account for 20% of the fatty acids in the bound fatty acid fraction and for 94% (1.07 mg/gTOC) of the free fatty acid fraction. In contrast, with 0.82 mg/gTOC, the n-C_{16} and n-C_{18} dominated short chain fatty acids account for 84% of the bound fatty acid fraction and with 0.07 mg/gTOC only for 6% of the free fatty acid fraction. This distribution difference is most obvious within the samples (G001995, G001996 and G001989) at the transition from the early catagenesis to the main catagenesis (due to the in chapter 6 discussed increase of n-C_{16} and n-C_{18} fatty acids during this stage), but it is also visible in most of the other samples. The percentage proportions are visualised in figure 7.10. Overall it can be concluded that in the kerogen-bound fatty acid fraction the short chain fatty acids represent about 53% and the long chain fatty acids about 39% of the total fatty acids, while in the bitumen fatty acid fraction the short chain fatty acids represent about 13% and the long chain fatty acids about 86% (Table 7.4).

This result is remarkable, because considering the fact that all fatty acids linked or free derived from the same contributing organic material forming the deposited organic matter in a sample during early diagenetic processes, it might be expected that the relative proportions of short and long chain fatty acids are similar in the bitumen and kerogen fraction. This

7.3. Results and discussion

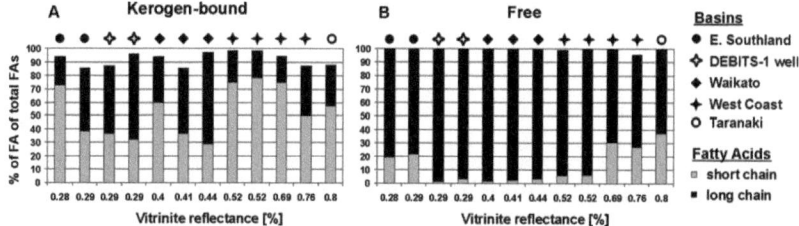

Figure 7.10.: *Percentage proportions in % of short chain fatty acids ($C_{13/14}$ to C_{19}) and long chain fatty acids (C_{20} to $C_{30/32}$) in the kerogen-bound fatty acid fraction (A) and the free fatty acid fraction (B).*

was found e.g. by Barakat and Rullkötter (1995) in samples from the sulfur-rich lacustrine sediments in the Nördlinger Ries (Germany). This would be even more true when considering the hypothesis that the organic matrix permanently releases former bound compounds into the surrounding being in a kind of an equilibrium. In Chapter 5, it was suggested that acidic hydrolysis may be a process that causes a constant release of low molecular weight organic acids (LMWOAs) into the surrounding sediments and, therefore, providing a permanent feedstock for microbial communities inhabiting the respective sediments. However, in contrast to the LMWOAs, the high molecular weight organic acids (HMWOAs) are only less or even not soluble in the ambient porewater and, thus, these fatty acids might not be in an equilibrium with the pore water and will not be removed from the system by porewater flow stimulating the balance reaction. Thus, a constant release by a hydrolysis balance reaction might not take place for the HMWOAs.

Nevertheless, if the source material is the same for the bound and free fatty acids, the differences in the relative proportions of short and long chain fatty acids in the free and the kerogen-bound fatty acid fractions are astonishing and the reasons for this can only be speculated on here. One reason might be that short chain fatty acid are preferentially removed from the bitumen fraction. Although HMWOAs have a low solubility in the porewater, the solubility of the short chain length fatty acids might be slightly higher

7. Free Fatty Acids, Alcohols And Esters From The Bitumen Fraction

than for the long chain fatty acids enabling the transport of the short chain fatty acids by pore water and improving their bioavailability for microbial degradation.

In organic carbon rich deposits, numerous microorganisms are involved in the formation of the complex organic matter during early diagenesis and dwell within the pores of the peat or lignite matrix. After cell death, the main part of their own organic matter might be directly incorporated into the organic macromolecular matrix, leading to a generally higher proportion of short chain fatty acids in the kerogen bound fraction and a smaller proportion in the bitumen fraction. In Chapter 6 it was already suggested that a part of n-C_{16} and n-C_{18} dominated short chain fatty acids may derive from bacterial membrane lipids. The observed increase of kerogen-bound n-C_{16} and n-C_{18} fatty acids with unusual immature CPI_{FA} values in samples from the early to main catagenesis (*cf.* Figure 6.5) was suggested to reflect the supply of microbial biomass to the kerogen by microorganisms stimulated by the slowly increasing substrate release due to the gradual onset of thermal cracking reactions in the kerogen. The fact that such an increase cannot be observed in the bitumen fatty acid fraction (Figure 7.3) together with the generally higher proportion of short chain fatty acids in the kerogen bound fraction supports the suggestion, that a significant proportion of the short chain fatty acids in the bitumen and kerogen bound fraction is of different origin. A microbial origin for these high amounts of short chain fatty acids is highly conceivable. Furthermore, another possibility might be that the microbial biomass is trapped within the macromolecular matrix and, therefore, can only insufficiently be extracted by simple organic solvent extraction.

Another aspect to compare is the relative proportion of the total free fatty acids to the total bound fatty acids as calculated in table 7.5. The bound fatty acids are considered to be less susceptible to diagenetic changes. Investigating short surface cores of 0 to 50 cm depth, Farrington and Quinn (1971, 1973) were able to show smaller variation of the bound fatty acid frac-

7.3. Results and discussion

Sample	Basin/Well	R_0 [%]	KFA [µg/gTOC]	BFA [µg/gTOC]	KFA+BFA [µg/gTOC]	KFA [%]	BFA [%]	$\frac{KFA}{KFA+BFA}$
G001978	E. Southland	0.28	602	710	1312	45.9	54.1	0.46
G001976	E. Southland	0.28	358	840	1198	29.9	70.1	0.3
G004544	DEBITS-1	0.29	2023	14250	16273	12.4	87.6	0.12
G004541	DEBITS-1	0.29	1277	11180	12457	10.3	89.7	0.1
G001982	Waiakto	0.4	788	7820	8608	9.2	90.8	0.09
G001983	Waikato	0.41	428	1870	2298	18.6	81.4	0.19
G001980	Waikato	0.44	387	830	1217	31.8	68.2	0.32
G001995	West Coast	0.52	771	1050	1821	42.3	57.7	0.42
G001996	West Coast	0.52	1049	1140	2189	47.9	52.1	0.48
G001989	West Coast	0.69	281	390	671	41.9	58.1	0.42
G001993	West Coast	0.76	265	220	485	54.6	45.4	0.55
G001991	Taranaki	0.8	137	240	377	36.3	63.7	0.36

Table 7.5.: *Amounts of all kerogen bound fatty acids (KFA) and bitumen fatty acids (BFA) and total amount of fatty acids (KFA + BFA), percentage proportions of KFA and BFA of the total fatty acids and the ratio of KFA to the sum of KFA + BFA indicating the dominance of bound or free fatty acids.*

tion with increasing depth than for the free fatty acid fraction. The bound fatty acids appear to be more resistant to diagenetic alteration and are suggested to become an increasingly larger portion of the total fatty acids in sediments with increasing burial depth.

With the exception of the two Eastern Southland Basin samples (G001978 and G001976), this trend can also be seen in Figure 7.11 although the current sample set covers a significantly broader maturity range. The parameter KFA/(KFA+BFA) was used to indicate the ratio of the kerogen-bound fatty acids (KFA) to the total fatty acids (sum of KFA + bitumen fatty acids (BFA)) obtained from each sample. As expected, the amount of kerogen-bound fatty acids increases during diagenesis and early catagenesis (Figure 7.11, Table 7.5) reflecting the lower susceptibility to diagenetic degradation during maturation. The two samples from the Eastern Southland Basin,

7. Free Fatty Acids, Alcohols And Esters From The Bitumen Fraction

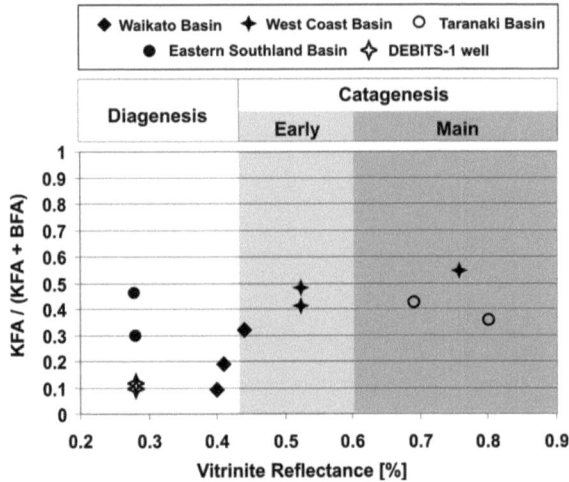

Figure 7.11.: *The proportion of the kerogen fatty acids (KFA) to the total fatty acids (kerogen fatty acids (KFA) + bitumen fatty acids (BFA) indicating the different susceptibility of free and kerogen-bound fatty acids to diagenetic processes.*

however, show an unexpected high KFA content. The reason for this can be seen in Figure 7.3. The amounts of the free fatty acids from these samples are extremely low, due to an unusually low amount of long chain fatty acids. This points to a significant lower plant wax contribution in these samples which might reflect variations in the facies of the contributing plant material. In contrast, the KFA obtained from these samples are comparatively high due to the dominating influence of the short chain fatty acids. The DEBITS-1 well samples (G004541 and G004544) being in the same thermal maturity range (0.29%), however, show the expected high amounts of BFAs and, therefore, their KFA/(KFA+BFA) ratio is low and plot within the expected trend (Figures 6.3, 7.3 and 7.11).

Vu (2008) were able to show that thermal degradation slowly starts between 0.4 and 0.5% R_0 and that the onset of thermal oil generation from the NZ coals is located at about R_0 0.6%. However, a significant shift again to a larger bitumen fraction cannot be observed in the parameter values, although a slight decrease might be indicated at the end of the maturity series. A reason for this might be that thermal degradation of the fatty acids linked to the kerogen might not lead to free fatty acids but to free hydrocarbons due to decarboxylation reaction. Thus, thermal cracking would not increase the free fatty acid fraction.

Alcohols

Free and ester-bound alcohols are both only present in significant amounts in the samples at an early diagenesis stage (*cf.* Figures 6.3, 6.4, and 7.6, Table 7.2). They show a rapid decrease during early maturation and almost disappeared from $R_0 > 0.3\%$. Free and kerogen bound alcohols show a similar distribution pattern with a maximum near or at $n\text{-}C_{24}$ (*cf.* Figures 6.4 and 7.6). The free alcohols are significantly less abundant than the ester-bound alcohols, e.g. for the Eastern Southland samples G001978 (free: 23.7 µg/gTOC vs. kerogen-bound: 1250 µg/gTOC) and G001976 (free: 94.4 µg/gTOC vs. kerogen-bound: 1485 µg/gTOC) (*cf.* Tables 6.1 and 7.2).

Functionalised pentacyclic triterpenes

The pentacyclic triterpenes from the bitumen fraction are totally different from those of the bound fraction. Betulin and lupenol were only found after saponification of the pre-extracted samples. Thus, they are directly incorporated into the macromolecular matrix. In contrast, betulin and lupenol were not present in the bitumen fraction.

In contrast, the triterpenes taraxerol, taraxerenone, oleanenone, ursenone, friedelin, β-amyrin, δ-amyrin and lupenone were only found in the bitumen fraction but they are absent in the saponification products and thus, they are probably not incorporated into the kerogen. Taraxerone, oleanone, ursenone and fridelin are ketones (Figure 7.9) and, therefore, it is obvious that they are not chemically bound via ester functions to the kerogen matrix. However, taraxerol as well as β- and δ-amyrin (Figure 7.9) are

7. Free Fatty Acids, Alcohols And Esters From The Bitumen Fraction

alcohols like lupeol and betulin (cf. Figure 6.9) but in contrast to the latter, they are not linked into the macromolecule. A reason for these differences remains unclear, but might be found in the original organic source material and in early diagenetic processes being responsible for the formation of the bitumen and kerogen.

7.4. Conclusion

The fatty acid and neutral polar fraction obtained from the bitumen of 12 lignite and coal samples of different maturity were analysed and compared with the corresponding fractions obtained from the kerogen after alkaline ester cleavage of the same samples as described in chapter 6. The bitumen fractions contained fatty acids, fatty acid methyl esters, alcohols and functionalised pentacyclic triterpenoids.

The concentrations of the bitumen short chain (n-C_{14} to n-C_{19}) and long chain (n-C_{20} to n-C_{32}) free fatty acids decrease with ongoing maturation. A main decrease can be observed from diagenesis to early catagenesis, and during the main catagenesis, only very low amounts of fatty acids are detectable. This decrease is even clearer if using the CPI_{FA} parameter, suggested to be less facies dependent and more maturity related. Fatty acid methyl esters (FAME) show a comparable picture and may derive from methylation of free fatty acids within the sediment or during the organic solvent extraction processed with methanol.

Interesting differences become evident when comparing the free and kerogen-bound fatty acids. Although the fatty acid distribution patterns are on a qualitative level similar, the relative proportion of the short chain fatty acids to long chain fatty acids is significantly higher in the kerogen-bound fraction. Thus, the bitumen and kerogen fraction seem not to be in a kind of an equilibrium. It is suggested that the higher proportion of short chain fatty acids in the kerogen are remains of microbial communities being

7.4. Conclusion

involved in the transformation of the organic carbon rich formation during ongoing maturation. A higher proportion of the microbial biomass might directly be incorporated into the kerogen matrix during this process after cell death. Especially, the increase in the n-C_{16} and n-C_{18} fatty acids, not visible in the bitumen faction, during the early catagenesis was related to a microbial origin. During this period, the geothermally-induced slow release of hydrocarbons started being a potential feedstock for deep microbial life.

Generally, the kerogen-bound fractions are suggested to be less susceptible to degradation than the bitumen fraction. A ratio of kerogen-bound fatty acids (KFA) versus the bitumen fatty acids (BFA) using the parameter KFA/(KFA+BFA) of the NZ coal series showed increasing values and revealed the expected trend, especially from the diagenesis to early catagenesis indicating a lower susceptibility to microbial degradation of the kerogen-bound fatty acids. During ongoing catagenesis, the ratio remains constant or might slightly decrease again, a reason for this being most likely the beginning thermal degradation of the kerogen.

As already observed in the fraction of kerogen-bound alcohols, the amount of free alcohols show a strong decrease during early diagenesis, pointing to a rapid degradation of the alcohols. Several functionalised pentacyclic triterpenes, mainly related to mangrove swamp plant distribution, were detected in the neutral polar fraction of the bitumen. These triterpenoids were not found within the ester-bound fraction, indicating that these terpenoids have not been incorporated into the matrix during diagenesis.

8. Organic Acids From Organic Matter Rich Layers From The DEBITS-1 Well

8.1. Introduction

Microorganisms represent the oldest and most widespread life form on Earth. They are able to effectively adapt to varying environmental conditions such as pressure, temperature and nutrient availability (Guckert et al., 1986; Russell and Fukunaga, 1990; Yano et al., 1998; Fang et al., 2000; Mangelsdorf et al., 2005). Therefore, they have been found to be present in many extreme environments such as deep marine hydrothermal vents (Jørgensen et al., 1992), hypersaline lakes (Jonkers et al., 2003), in polar- and permafrost areas (Gilichinsky and Wagener, 1995; Panikov and Sizova, 2007), in the deep sea as well as in deep terrestrial sediments (Fredrickson and Onstott, 1996; Onstott et al., 1997). Microorganisms were also found in oil reservoirs with temperatures below 80°C utilising hydrodcarbons as nutrients. The biodegradation of oil leads to a decrease in paraffin content and an increase in oil density, sulphur content, acidity and viscosity with negative economic consequences for oil production (ZoBell, 1945; Stetter et al., 1993; Rueter et al., 1994; Wilhelms et al., 2001). With the improvement in the development of analytical detection techniques, the widespread occurrence of diverse microbial communities has been demon-

8. Organic Acids From Organic Matter Rich Layers From The DEBITS-1 Well

strated (Parkes et al., 1994; L'Haridon et al., 1995; Krumholz, 2000; Pedersen, 2000; D'Hondt et al., 2004) and as a result of these findings, the great importance of the subsurface biosphere contributing to all elemental cycles is now accepted in the scientific community (Whiteman et al., 1998).

However, the diverse mechanisms by which these microbial communities interact with the geosphere are still poorly understood (Lovley and Chapelle, 1995; Horsfield et al., 2006, 2007). One of the very important questions in this context is the search for substrates providing carbon and energy sources for these deep microbial communities, in order to drive their metabolic cycles.

In subsiduous sedimentary basins, buried organic matter is thought to be the most likely feedstock for the deeply buried microorganisms (Wellsbury et al., 1997; Horsfield et al., 2006; Parkes et al., 2007). During maturation the buried organic matter undergoes diverse geochemical alteration processes, a part of which is the loss of oxygen-containing compounds, as indicated by the increase of the C/O atomic ratio of the kerogen, mainly during diagenesis and early catagenesis. Thus, organic carbon rich layers (e.g. coal seams) may form a potential feedstock for microbial ecosystems in deep sedimentary successions. Another important parameter that regulates microbial activity within the sediments was shown by Fredrickson et al. (1997). Investigating a shale-sandstone sequence from northwestern New Mexico, they were able to show that the pore throat size within the rocks may regulate the observed level of microbial activity. As a result of this study, they suggested that growth and metabolism of shale-bound organisms may be limited by slow diffusion of nutrients or by the inability of microbes to migrate through the pore throats.

Sedimentary sequences obtained from the DEBITS-1 well represent an excellent opportunity to investigate the interaction between feedstock lithologies and microbial ecosystems. The DEBITS-1 well penetrates sequences containing organic carbon rich layers intercalated by organic carbon lean silt and sandstones and some claystones showing moderate organic carbon contents. The coarser grained silt and sandstones with sufficient pore throats close to the organic carbon rich deposits might act as life habitats for microbial communities (Krumholz, 2000). Low molecular weight organic acids such as formate, acetate and oxalate represent important substrates for microbial metabolism (Sansone and Martens, 1981, 1982; Sørensen et al., 1981; Jørgensen, 1982). Thus, kerogen bound LMWOAs form a potential feedstock for deep deep microbial life (Glombitza et al., 2009b).

To investigate the content of ester-bound LMWOAs in lignites and coals and in the intercalated clay-, silt- and sandstones, an alkaline ester cleavage approach was applied that was especially designed to analysed LMWOAs linked to the kerogen matrix (*cf.* Chapter 5 (Glombitza et al., 2009b)). In addition, the distribution of high molecular weight ester-bound fatty acids and alcohols in lignite and coal layers from the DEBITS-1 well was investigated to reveal changes in the content of oxygen containing compounds with increasing burial depth and, therefore, increasing maturation using the ester cleavage procedure described by Höld et al. (1998).

8.2. Samples and methods

8.2.1. Sample material

For this study, 16 samples from the 148 m deep DEBITS-1 well were chosen (*cf.* Chapter 3.1.2). The DEBITS-1 well was drilled on an open ground near the small village of Huntly, located in the Waikato coal area, on the North Island of New Zealand. Ten of the selected samples belong to different organic matter rich lithologies penetrated by the DEBITS-1 drill core. The stratigraphy covered by the DEBITS-1 well is shown in Figure 3.2 (Chapter 3). Seven samples from above the unconformity at ca. 76 m depth (Tauranga Group) are lignites with a thermal maturity of 0.29% vitrinite reflectance, including two samples from subcore 6.3 (ca. 19 m), two samples from subcore

8. Organic Acids From Organic Matter Rich Layers From The DEBITS-1 Well

11.2 (ca. 31 m), one sample from subcore 17.4 (ca. 63 m) and two samples from subcore 18.2 (ca. 65 m). The remaining three samples were from sub-bituminous coal layers below the unconformity (Te Kuiti Group) with a vitrinite reflectance of 0.39%, including two samples from subcore 29.5 (ca. 130 m) and one from subcore 32.2 (ca. 142 m). All lignite and coal samples are in the diagenetic maturity stage. The Tauranga Goup lignites represent early diagenesis, while the Te Kuiti Group coals represent late diagenesis.

Additionally, a series of six samples from subcore 9.2 (ca. 26 m) were selected representing a transect from sandstone, siltstone and claystone to lignite, to analyse the distribution of microbial communities in relation to the potential feedstock (ester-bound LMWOAs) provided by the lithological system. Further information about the sample material can be found in Table 3.3.

8.2.2. Sample preparation

All samples were freeze dried and grounded before extracted with water and organic solvents, as described in chapter 4. Samples from subcore 9.2 were treated according to the method described in Chapter 4.1.4 and 5.3 for the analysis of ester-bound LMWOAs.

The organic matter (OM) rich samples of subcores 6.3, 11.2, 17.2, 17.4, 18.2, 29.5 and 32.2 were also analysed for the distribution of LMWOAs and additionally for the distribution of ester-bound high molecular weight organic acids (HMWOAs) and alcohols as described in Chapters 4.1.4 and 6.

8.3. Results and discussion

First the results for the lignite and coals are presented (Chapters 8.3.1 and 8.3.2) and then the relationship to other lithologies given in the context of

8.3. Results and discussion

potential habitats and available feedstocks for the deep biosphere (Chapter 8.3.3).

8.3.1. Low molecular weight organic acids from lignite and coal layers of the DEBITS-1 well

To trace changes of the ester-bound potential substrates for deep microbial life (e.g. LMWOAs) with burial depth, 10 samples from different organic matter rich lithologies penetrated by the DEBITS-1 well were chosen and analysed using the alkaline ester cleavage approach described in Chapter 5 (Glombitza et al., 2009b). Additionally, the results from the lignite sample G004535 from subcore 9.2 (see below) were added to the data set.

The released LMWOAs from the organic matter rich layers were formic acid, acetic acid and oxalic acid. In contrast to the ion chromatography (IC) analysis of the coal mine samples reported in chapter 5 (Glombitza et al., 2009b), in this study, also propionic acid and butyric acid were detected (Table 8.1, Figure 8.1). Their amounts are generally very low, compared to formic acid, acetic acid and oxalic acid. The reason that these compounds could be detected in this study could be the result of the application of a new suppressor and a new conductivity detection cell improving the resolution and sensitivity of the IC instrument. (Figure 8.1). The lignite samples from the Tauranga Group (R_0 0.29%, early diagenesis) above the unconformity at 76 m depth (subcores 6.3, 9.2, 11.2, 17.4 and 18.2) generally show a higher abundance of LMWOAs, whereas the sub-bituminous coal samples from the Te Kuiti Group (R_0 0.39%, late diagenesis) below the unconformity (subcores 29.5 and 32.2) revealed significantly lower amounts of these compounds (Figure 8.1, Table 8.1).

8. Organic Acids From Organic Matter Rich Layers From The DEBITS-1 Well

Sample	Subcore	Lithology	TOC [%]	Mean Depth [m]	Formate [mg/gTOC]	Acetate [mg/gTOC]	Propionate [mg/gTOC]	Butyrate [mg/gTOC]	Oxalate [mg/gTOC]
G004541	6.3	lignite	45.11	18.94	17.48	3.88	-	-	3.88
G002813	6.3	lignite	40.00	19.36	10.04	1.85	0.06	0.29	0.48
G004535	9.2	lignite	18.50	26.16	5.82	2.85	0.14	0.41	3.40
G003579	11.2	lignite	37.16	31.24	7.83	2.08	-	0.25	5.87
G003580	11.2	lignite	32.81	31.77	17.53	2.31	0.12	0.51	0.98
G004544	17.4	lignite	37.87	63.46	28.65	3.25	-	-	5.72
G003577	18.2	lignite	41.79	65.40	8.24	2.06	-	0.28	10.94
G003578	18.2	lignite	40.48	65.54	5.18	1.89	-	0.22	7.87
G003575	29.5	sub-bit. coal	58.29	130.53	3.39	1.67	0.09	0.14	0.87
G003574	29.5	sub-bit. coal	57.72	130.83	2.49	1.49	0.08	0.13	0.88
G003576	32.2	sub-bit. coal	59.59	142.52	0.44	0.25	0.01	0.02	0.13

Table 8.1.: *LMWOAs released from lignite and sub-bituminous samples of DEBITS-1 well subcores 6.3, 9.2, 11.1, 17.4, 18.2, 29.5 and 32.2 by alkaline ester cleavage reaction in mg/gTOC.*

8.3. Results and discussion

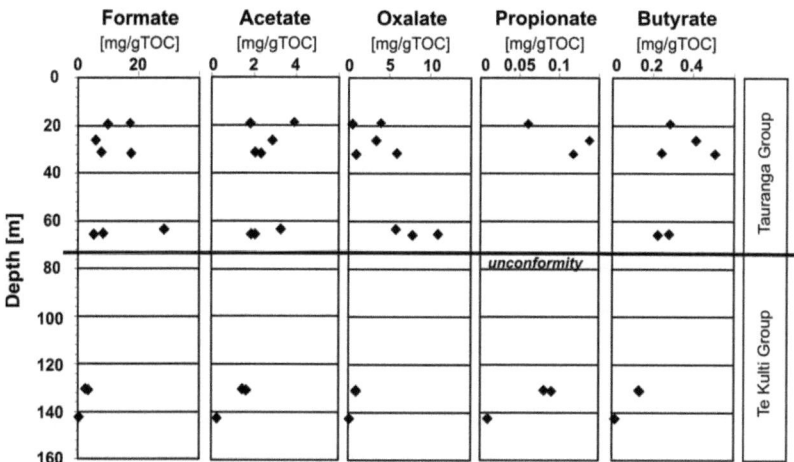

Figure 8.1.: *Ester-bound LMWOAs from lignite and sub-bituminous coal layers of the DEBITS-1 well. (Please note different scales of x-axis.)*

Generally, the most abundant acid is formic acid with up to approximately 28 mg/gTOC in the Tauranga Group lignites and up to 3.4 mg/gTOC for Te Kuiti Group coals followed by oxalic acid with up to approximately 11 mg/gTOC in the Tauranga Group lignites and up to 0.9 mg/gTOC in the Te Kuiti Group coals and acetic acid with up to 4 mg/gTOC in the Tauranga Group lignites and up to 1.7 mg/gTOC in the Te Kuiti Group coals. Propionic acid and butyric acid are only present in minor amounts (Tauranga Group: propionic acid up to 0.1 mg/gTOC, butyric acid up to 0.5 mg/gTOC; Te Kuiti Group: propionic acid up to 0.09 mg/gTOC, butyric acid up to 0.14 mg/gTOC). Propionic and butyric acids were not detected in all samples, probably due to their generally lower abundance which might be close to the detection limit of the applied analytical procedure.

These findings revealed that all organic matter rich lithologies in the Tauranga Group as well as in the underlying Te Kuiti Group contain notable amounts of LMWOAs bound via ester bonds to the kerogen matrix. Analysing the free (unbound) organic acids in the water extract of the lignite and coal samples from the DEBITS-1 well Vieth et al. (2008) also showed the presence of formic, acetic and oxalic acid decreasing with ongoing maturation. Highest amounts were found in the Tauranga Group water extract samples and samples from the Te Kuiti Group contained significantly lower but still detectable amounts of these LMWOAs. Comparing the free (Vieth et al., 2008) and kerogen-bound LMWOAs from the organic matter rich samples of the DEBITS-1 well it can be seen that samples with high amounts of kerogen-bound LMWOAs contain also high amounts of free LMWOAs being directly available for microbial metabolism. In contrast, samples with low amounts of bound LMWOAs also show low concentrations of free LMWOAs in the water extract. These results points to an potential equilibrium between kerogen-bound LMWOAs and those being in a free form which might be mediated by hydrolysis within the porewater supported by the good solubility of LMWOAs in the solvent water. Therefore, the organic matter rich lithologies are suggested to provide a substrate feedstock for deep microbial communities that is going to be released during further maturation processes.

8.3.2. Fatty acids and alcohols from lignite and coal layers of the DEBITS-1 well

The 10 lignite and coal samples from the Tauranga and Te Kuiti Group were additionally investigated for their content of ester-bound high molecular weight organic acids (HMWOA) and alcohols using the procedure reported by Höld et al. (1998) with some modifications (*cf.* Chapter 6 (Glombitza et al., 2009a)).

8.3. Results and discussion

Sample	Mean Depth [m]	Subcore	Total FAs [µg/gTOC]	Sum FAs (C_6-C_{12}) [µg/gTOC]	Sum FAs (C_{13}-C_{19}) [µg/gTOC]	Sum FAs (C_{20}-C_{30}) [µg/gTOC]	CPI_{FA} (C_{14}-C_{18})	CPI_{FA} (C_{20}-C_{30})	Total Alc [µg/gTOC]	CPI_{OH} (C_{18}-C_{30})
G004541	18.94	6.3	1276.98	8.12	411.72	812.51	27.76	10.63	2138.71	23.24
G002813	19.34	6.3	2130.78	2.91	269.81	1792.28	16.76	15.39	16037.39	12.23
G003579	31.24	11.2	953.73	4.22	224.31	655.06	14.57	9.02	1040.51	9.38
G003580	31.77	11.2	2853.72	13.79	552.14	1998.51	14.06	9.33	2122.31	9.2
G004544	63.46	17.4	2023.46	25.94	743.41	1015.99	29.64	4.25	61275.02	20.17
G003577	65.40	18.2	879.72	4.56	289.86	528.35	15.65	5.50	2427.79	7.21
G003578	5.54	18.2	533.46	1.90	229.37	249.65	17.38	4.95	477.85	6.53
G003575	130.68	29.5	281.00	-	68.71	222.06	10.93	4.05	-	-
G003574	130.3	29.5	462.84	3.60	258.32	236.59	12.57	5.01	27.29	10.37
G003576	142.66	32.2	261.55	1.63	176.03	111.45	11.10	5.11	19.86	5.85

Table 8.2.: *Fatty acids and alcohols released from organic matter rich layers of DEBIRT-1 well and calculated CPI for released fatty acids and alcohols.*

8. Organic Acids From Organic Matter Rich Layers From The DEBITS-1 Well

The fatty acids released by alkaline ester cleavage procedure were detected in the range of n-C_{13} to n-C_{32}. They revealed a bimodal distribution with a first maximum at n-C_{18} and a second maximum at n-C_{26} (Table 8.2). Ester bound alcohols were found in the range of n-C_{14} to n-C_{32} with a maximum at n-C_{26}. The amounts of the released fatty acids (FA) are generally higher in the upper Tauranga Group samples (Table 8.2) reflecting ongoing degradation and maturation of the organic matter with depth. A depth profile of the released fatty acids and alcohols from the DEBITS-1 well is shown in figure 8.2. Although a general decreasing trend can be observed for the total amounts of fatty acids, the values reveal a broad scattering, especially in the low maturity lignite samples of the Tauranga Group most likely due to different organic matter facies.

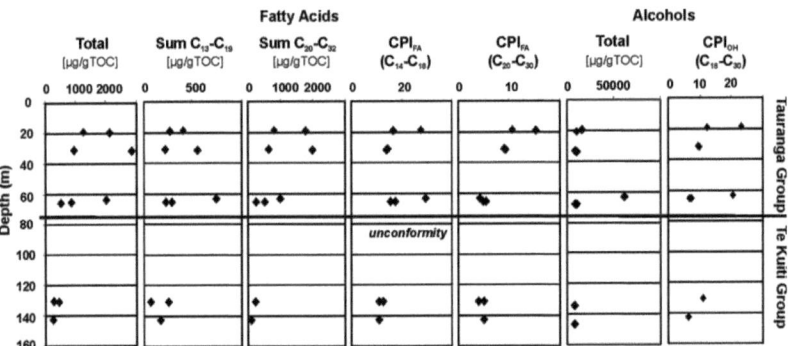

Figure 8.2.: *Ester-bound total as well as short chain (C_{13}-C_{19}) and long chain (C_{20}-C_{32}) fatty acids and total alcohols from lignite and coal layers of the DEBITS-1 well and carbon preference index data of short chain and long chain fatty acids and alcohols.*

The fatty acids were separated into short chain fatty acids (n-C_{13} to n-C_{19}) and long chain fatty acids (n-C_{20} to n-C_{32}) due to their different origin as already reported in chapter 6. Although the long chain fatty acids (n-C_{20} to n-C_{32}) deriving from terrestrial plant material show some variation in the Tauranga Group samples, generally, their amounts decrease with increasing burial depth. Because their abundances is up to four times higher than

8.3. Results and discussion

that of the short chain fatty acids, their amounts and trend appear to cause the overall decrease in the total fatty acids. Such an overall decrease could not be observed for the short chain fatty acids. The main compound determining the signal of the short chain fatty acids are palmitic acid and stearic acids. Especially sample G003580 (subcore 11.2, 31.77 m) and G004454 (subcore 17.4, 63,46 m) show high amounts of these fatty acids (Figure 8.2, Table 8.2). The distribution of both the long and the short chain fatty acids can be influenced by facies variation of the organic source material during deposition causing the observed variations in the Tauranga Group samples.

To obtain a maturation related and less facies dependent signal, the carbon preference index for fatty acids (CPI_{FA}) was applied for both carbon number ranges (Table 8.2, Figure 8.2) as introduced in Glombitza et al. (2009b). For the long chain fatty acids, this index revealed a strong decreasing trend with ongoing burial depth. Although there is no difference in the vitrinite reflectance data for the lignites of different burial depth from the Tauranga Group samples the strongly decreasing CPI_{FA} data outline a maturation related trend (Figure 8.2). This suggests that the CPI_{FA} of the terrestrial plant deriving fatty acids (n-C_{20}-n-C_{30}) represents a sensitive maturity parameter for lignites and coals.

However, this clear decreasing trend of the CPI_{FA} was not observed for the short chain fatty acids of the Tauranga Group (Figure 8.2). Although, a slight decrease of this CPI_{FA} data can be noticed and samples of the Te Kuiti Group show a significant lower CPI_{FA}, still a significant scattering can be observed. Therefore, it may be speculated that the maturation signal of the short chain fatty acids is affected by further influencing factors such as variable contributions of bacterial biomass containing palmitic acid and stearic acid as main cell membrane component (*cf.* Chapter 6, (Glombitza et al., 2009a)).

In addition, a CPI_{OH} was calculated using the amounts of released alcohols from C_{17} to C_{31} (Equation 8.1, Figure 8.2, Table 8.2).

8. Organic Acids From Organic Matter Rich Layers From The DEBITS-1 Well

$$CPI_{OH}(C_{18} - C_{30}) = \frac{1}{2}\left(\frac{\sum_{n=9}^{15} C_{2n}}{\sum_{n=9}^{15} C_{2n-1}} + \frac{\sum_{n=9}^{15} C_{2n}}{\sum_{n=9}^{15} C_{2n+1}}\right) \qquad (8.1)$$

The calculated CPI_{OH} also revealed a decreasing trend, but scattering is still present. Therefore, it seems that the CPI_{OH} is less usable as a maturation parameter in contrast to the CPI_{FA} for the terrestrial plant deriving fatty acids.

8.3.3. Feeding potential of organic carbon rich lithologies for microbial life in adjacent organic carbon poor layers from the DEBITS-1 well, subcore 9.2 (25.93-26.15 m)

To investigate the feeding potential of organic carbon rich lithologies for deep microbial communities samples from the DEBITS-1 well subcore 9.2 were selected. Subcore 9.2 covers a transect from organic carbon poor to organic carbon rich lithologies. From this transect, 6 samples of different lithologies were investigated. The uppermost sample of this transect is a sandstone (G004530), followed by a silty sandstone sample (G004531), a silt sample (G004532), a silty claystone sample (G004533), a sample containing clay and lignite (G004534) and the lowest sample being a lignite (G004535). The TOC values increase with increasing sample depth from 0.15% (G004530) to 18% (G004535) indicating the increasing content of organic matter towards the lignite layer (cf. Table 3.3).

These samples were treated by the alkaline ester cleavage approach developed in this thesis (cf. Chapter 5) to analyse the contend of ester-bound low molecular weight organic acids (LMWOAs). The main acids found within the samples are again formic acid, acetic acid, propionic acid, butyric acid and oxalic acid.

The content of LMWOAs increase with increasing TOC content and, therefore, with decreasing distance from the lignite layer (Figure 8.3, Ta-

8.3. Results and discussion

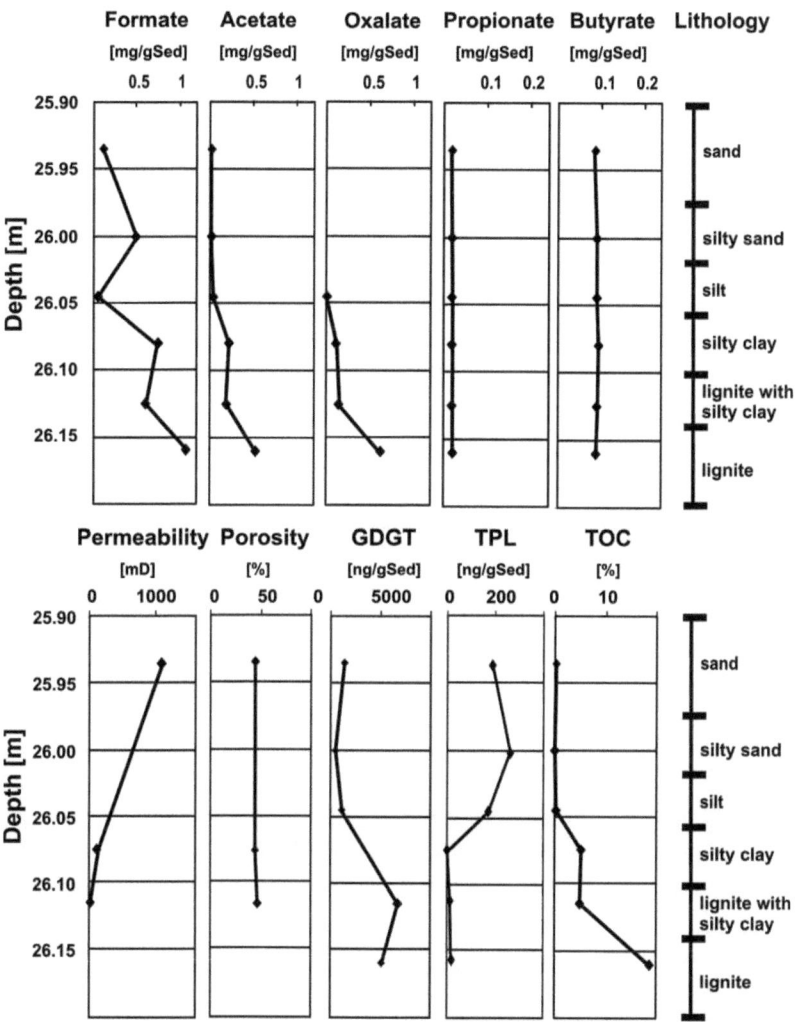

Figure 8.3.: *Ester-bound LMWOA from a sandstone, siltstone, claystone and lignite transect in subcore 9.2 (25.93 - 26.15 m), compared with glycerol dialkyl glycerol tetraethers (GDGT), total phospholipids (TPL), porosity, permeability and TOC. Porosity and permeability data were provided by Steven Franks (Saudi Aramco, Saudi Arabia), GDGT, TPL and TOC data were provided by K.Mangelsdorf (GFZ Potsdam, Germany). Please note the different scales.*

ble 8.3). The main LMWOA released is formic acid showing values of about 1 mg per g sediment in the lignite sample and even the OM poor sandstone sample released about 0.1 mg formic acid per g sediment. Acetic acid was found in lower amounts than formic acid. The highest content of acetic acid was released from the lignite layer sample (0.5 mg/gSed). The acetic acid content decreases with decreasing TOC content of the samples to 0.02 mg/gSed in the sandstone sample. The released oxalic acid shows the same decreasing trend but was only detected in four samples of subcore 9.2. The two samples with the highest distance from the lignite layer did not reveal any oxalic acid. This might be explained by the fact that oxalic acid is a common LMWOA especially in lignites and coals (Kawamura and Kaplan, 1987; Barth et al., 1988; Bou-Raad et al., 2000; Vieth et al., 2008). Propionic acid only revealed a slight decreasing trend from OM rich samples to OM poor samples. For butyric acids this trend could not be observed (Table 8.3). However, in this context, it has to be noted that the amounts of propionic and butyric acid are significantly lower than those of the other released LMWOAs (in the range of 0.01 - 0.1 mg/gSed, Table 8.3). These values were close to the detection limit of the applied total procedure and, therefore, their concentration variability might not be very significant.

In a similar transect sample set from the DEBITS-1 well (subcore 6.1, 18.27 m to 18.64 m), Vieth et al. (2008) analysed free LMWOAs obtained by water extraction of the samples. The extracted LMWOAs represent the LMWOAs which are directly available for potential microbial metabolism. In their study, Vieth et al. (2008) were also able to extract the highest amount of free LMWOAs from the sample with the highest TOC content, and the extraction yields decreased with decreasing TOC content coinciding with the distribution of the kerogen-bound LMWOAs. This points to a correlation of free and ester-bound LMWOAs and may indicate that former ester-bound LMWOAs were liberated during ongoing maturation maybe in a kind of an equilibrium as suggested in Glombitza et al. (2009b).

8.3. Results and discussion

Sample	Subcore	Lithology	TOC [%]	Mean Depth [m]	Formate [mg/gSed]	Acetate [mg/gSed]	Propionate [mg/gSed]	Butyrate [mg/gSed]	Oxalate [mg/gSed]
G004530	9.2	sandstone	0.15	25.94	0.12	0.02	0.02	0.06	-
G004531	9.2	silty sand	0.12	26.00	0.50	0.02	0.02	0.09	-
G004532	9.2	silt	0.28	26.05	0.06	0.04	0.02	0.08	0.01
G004533	9.2	silty clay	5.23	26.08	0.76	0.23	0.02	0.1	0.11
G004534	9.2	lignite/silty clay	4.86	26.13	0.61	0.19	0.02	0.08	0.15
G004535	9.2	lignite	18.50	26.16	1.08	0.53	0.03	0.08	0.63

Sample	Subcore	Lithology	TOC [%]	Mean Depth [m]	TPL [ng/gSed]	GDGT [ng/gSed]	Permeability [mD]	Porosity [%]
G004530	9.2	sandstone	0.15	25.94	185.96	1347.47	1020.00	44.61
G004531	9.2	silty sand	0.12	26.00	262.86	477.18	-	-
G004532	9.2	silt	0.28	26.05	170.86	1108.78	-	-
G004533	9.2	silty clay	5.23	26.08	-	-	118	45.09
G004534	9.2	lignite/silty clay	4.86	26.13	-	6775.19	20.70	47.42
G004535	9.2	lignite	18.50	26.16	-	5026.34	-	-

Table 8.3.: *Low molecular weight organic acids released from samples of a sand to lignite transsect of the DEBITS-1 well of 9.2 obtained after alkaline ester cleavage reaction in mg/gSed. Additionally listed are the data for the glyceryl dialkyl glycerol tetraethers (GDGT's), total phospholipids (TPL), porosity and permeability, provided by K. Mangelsdorf (GFZ Potsdam).*

8. Organic Acids From Organic Matter Rich Layers From The DEBITS-1 Well

In order to compare the distribution pattern of potential substrates detected within the sediment samples of different maturity with indicators for present and past microbial activity, the occurrence of specific microbial biomarkers are shown in figure 8.3 (Table 8.3). The presence of phospholipid (PLs) deriving from bacterial cell membrane are used to indicate the occurrence of viable microbial ecosystems due to the fact that these biomarkers are relatively unstable rapidly degrading after cell death (Harvey et al., 1986; White et al., 1979; Rütters et al., 2002; Zink et al., 2003; Sturt et al., 2004). Therefore, they are also called "life markers". Core tetraether lipids (glycerol dialkyl glycerol tetraethers, GDGTs) with isoprenoidal bridges derive from the mono layered cell membranes of archaea and those with only branched GDGT from bacteria (Hopmans et al., 2004; Weijers et al., 2006; Yamamoto et al., 2008). These tetraether lipids have already lost their head groups and are, therefore, already partly degraded. However, their cores are more stable with respect to degradation and can, therefore, be used as paleo lifemarkers.

In Figure 8.3 it can be seen that the total phospholipids (TPL) are most abundant in the sandstone and silty sandstone samples adjacent to the lignite layer but they decrease in the silt sample and were not detected in the clay and lignite samples, although the substrate pool in form of LMWOA show its highest abundance in the clay and lignite layers. This appears on a first view contradictory. A reason for this may be the available proportion of interconnected pore space of the sediments that microbes used as their life habitats and which is suggested to support microbial exchange processes (Fredrickson et al., 1997). The porosity data shown in figure 8.3 indicate no significant variations between the sand-, silt- and claystones, however, the permeability being a measure for the connectivity of the pores significantly decreases from the sand- and siltstones to the claystones. The type and degree of interconnetion of the pores is a controlling factor for the occurrence of the deep biosphere because microorganisms occupy only about one millionth of available porespace (Parkes et al., 2000; Horsfield et al., 2007). Porosity and permeability of the lignite sample could not be determined but is suggested to be comparable to that of the claystone. These data suggest that a viable microbial biosphere is associated to the organic carbon rich lithologies in the adjacent organic carbon poor but more permeable lithologies. These microbial communities seem to be located in a zone with sufficiently high permeability close to an organic carbon feeder lithology being able to release appropriate amounts of substrates such as LMWOAs into the coarser grained adjacent sediments.

An analogue result was presented by Krumholz et al. (1997) by investigating the activity of sulfate reducing bacteria in an interface of highly permeable organic-poor sandstones and low permeable organic-rich shales at the Cerro Negro site, near Seboyeta (New Mexico, USA). They observed the highest microbial activity in the sandstone samples from near the sandstone/shale interface and correlated the observed microbial activity with the TOC depth distribution. The highest TOC contents were located in the shales close to the zones of high microbial activity in the TOC poor sandstones. They suggested that the organic matter trapped in the shales during deposition was diffusing across the shale/sandstone interface and thereby feeding the microbial community within the adjacent sandstone.

At first sight, these findings seem to contradict the results discussed in chapter 6 where high amounts of short chain fatty acids with an immature CPI_{FA} signature in the coals from the early catagenesis were suggested to be from microorganisms stimulated by the geothermally induced release of hydrocarbons. However, it must be mentioned that test experiments of phospholipids mixed with different amounts of organic matter matrix showed that the detection of phospholipids is significantly impeded with increasing TOC content of the background matrix (Mangelsdorf, unpublished data). This is because in samples with high TOC contents there is a cumulative competitional behavior of an increasing number of ions in the electrospray interface of the HPLC-ESI-MS system used for the detection of phospholipids (Mangelsdorf, unpublished data). Furthermore, the coals have been uplifed and contain cleats with higher permeability where microorganisms might

8. *Organic Acids From Organic Matter Rich Layers From The DEBITS-1 Well*

easily reside. Anyway, it is hard to believe that microorganisms should not also inhabit the organic carbon rich lithologies because they are overburden with potential substrates. Indeed, Krumholz et al. (1997) showed higher microbial activity adjacent to the TOC rich shales, but they also showed although to a lower extent microbial activity in the shales indicating that microorganisms also occupy the organic carbon rich lithologies. The same might be seen here in subcore 9.2. A larger microbial community dwells associated to the lignite layer in the adjacent silt and sandstone and a smaller population within the lignite layer which is obscured by the detection problem of phospholipids in TOC rich lithologies.

In contrast to the life marker PLs, the GDGTs detected in the lignite to sandstone transect of the DEBITS-1 well reveal their highest abundance in the claystone and lignite and are significantly lower in the adjacent coarser grained lithologies. This suggests that the GDGTs represent past microbial communities during time of sedimentation.

8.4. Conclusion

In this study, samples from organic matter rich layer of the 148 m deep DEBITS-1 well were analysed for their content of ester-bound LMWOAs as well as HMWOAs and alcohols. The samples of the Taugranga Group OM rich sediments (0 to 76 m) representing early diagenesis stage lignites (R_0 0.29%) contained high amounts of kerogen-bound formate, acetate, oxalate as well as propionate and butyrate. The sub-bituminous coal samples from the underlying Te Kuiti group representing late diagenesis maturity stage (R_0 0.39%) contained significanly lower but still notable amounts of these LMWOAs. Thus, a maturity related decrease of these compounds is suggested which is in accordance with the already described release of LMWOAs from OM rich lithologies during ongoing maturation processes (Glombitza et al., 2009b). Therefore, these OM rich layers appear to provide a significant feedstock potential for deep microbial life.

8.4. Conclusion

The samples were additionally investigated for their content in HMWOAs and alcohols. It was shown that the Tauranga Group lignites contain high amounts of short (n-C_{13} to n-C_{19}) and long (n-C_{20} to n-C_{32}) chain fatty acids with a broad scattering that is presumably facies related. At least for the long chain fatty acids the calculated CPI_{FA} values, a parameter suggested to be less facies dependent, show a clear decreasing trend pointing to ongoing maturation of the OM, although all Tauranga Group lignites have the same thermal maturity as indicated by the vitrinite reflectance data. The CPI_{FA} of the short chain fatty acids, however, still reveals some scattering which suggests that the short chain fatty acids are more affected by non maturity related factors like for example the contribution of short chain fatty acids from deep microbial sources. According to the ongoing maturation, the compound concentrations as well as the CPI_{FA} values from the samples of the underlying more mature Te Kuiti Group are somewhat lower and less variable, being in accordance with the higher level of maturation.

A transect (DEBITS-1 well subcore 9.2, 25.93-16.15 m) from organic carbon rich (lignite) to organic carbon poor lithologies (sandstone) was selected to investigate the feeding potential of these layers for associated microbial communities. Phospholipids (indicators for viable microbial communities) show high abundances in the more permeable TOC poor sand- and siltstones adjacent to the organic carbon rich lithologies (claystone and lignite). This suggests that these microbial communities are fed by substrates being released from the organic carbon rich lithologies into the surrounding sediments. Concentrations of GDGTs (biomarkers for paleo microbial communities) decrease from the organic carbon rich litholgies to the organic carbon poor adjacent strata. This seems to indicate the quantitative distribution of the microbial communities during time of sedimentation with respect to the deposited organic matter.

Although PLs were not detected in the claystone and lignite samples, the occurrence of microorganisms are also assumed in these lithologies, because of the high excess of substrates. However, their activity and maybe also their abundance might be higher in the more permeable lithologies as shown at least for the activity by Krumholz et al. (1997). A reason why PLs were not detected is given by the observation that a high background signal of organic matter can suppress the PL signals in the electrospray (ESI) interface during sample measurement on a HPLC-ESI-MS system.

9. Kinetics Of The Hydrolysis Of Low Molecular Weight Organic Acids

9.1. Introduction

In two of the previous Chapters (5 and 8), the cleavage of ester-bound low molecular weight organic acids (LMWOAs), such as formic acid, acetic acid and oxalic acid from coals and lignites were described and their potential to sustain microbial life within the deep Earth was estimated. For the cleavage of ester bonds in the kerogen samples, alkaline hydrolysis (saponification) was used. The alkaline hydrolysis represents a relatively fast and effective ester cleavage reaction and was, therefore, used to estimate the overall potential of the sediments to release LMWOAs. In contrast, the pore water of coaly layers is generally more in the neutral to slightly acidic pH ranges (pore water pH values between 6 and 8 were measured for the DEBITS-1 well samples; Vieth et al., 2008). Thus, the natural abiotic cleavage pathway of ester bond in lignites or coals is most likely proton-catalysed aqueous hydrolysis.

The processes leading to the abiotic cleavage of ester bonds in natural environments are suggested to be either a thermal radical cracking process which requires high temperatures or, at lower temperatures, hydrolysis within the pore water system (Eglinton et al., 1991; Siskin and Katritzky,

9. Kinetics Of The Hydrolysis Of Low Molecular Weight Organic Acids

1991). Increasing temperature and pressure during burial can effectively accelerate the hydrolysis process, even at neutral pore water pH values (Siskin and Katritzky, 1991). For the diagenetic to early catagenetic maturity range hydrolysis is assumed to be the most favorable abiotic process being responsible for the cleavage of ester bonds within the sediments (Siskin and Katritzky, 1991). Furthermore, free and water soluble acids (such as formic or acetic acid) being the products of ester cleavage reaction are suggested to autocatalyse the cleavage reaction (Siskin and Katritzky, 1991). Investigation of the kinetic parameters of this processes could help to evaluate the importance of the hydrolytic release of LMWOAs from the sediments. To estimate whether the hydrolytic clelavage of kerogen-bound LMWOAs might play a role in regulating the available feedstock for microbial communities two things are of importance: (1) The reaction rates of the hydrolysis generating the free LMWOAs and (2) the transport of the substrates within the pore water depending on the permeability of the sediment.

Thermal cracking reactions are unimolecular reactions (Atkins, 1986) and, therefore, the reaction rate only depends on the concentration of the starting molecule (the ester in case of ester cleavage process). The reaction rates can be described by first order reaction kinetics. The hydrolysis reaction is a bimolecular reaction (Day and Ingold, 1941), but due to the excess of water compared with the amounts of esters, the concentration of water can be assumed as constant, because it will not significantly change during hydrolysis. Thus, the reaction is a so-called pseudo first order reaction and can also be described by first order reaction kinetics (Atkins, 1986).

In this chapter, kinetic investigation of the acid-(proton)-catalysed hydrolysis reaction releasing LMWOAs from organic matter (OM) rich sediments is described. A series of NZ coal samples being in different thermal maturation stages from late diagenesis to main catagenesis was selected. First order reaction kinetic was assumed for the calculation of the kinetic parameters such as the activation enery (E_A), the frequency factor (A), the reaction rate constant (k) and the half life ($t_{1/2}$).

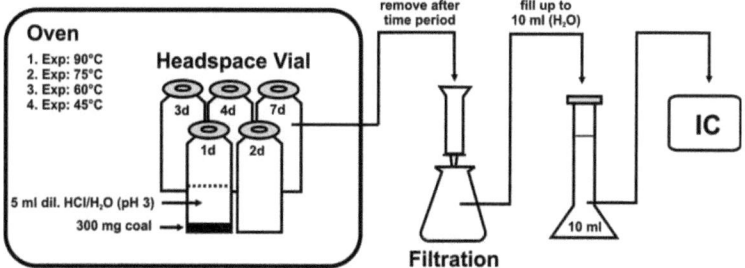

Figure 9.1.: *Experimental procedure to investigate the kinetics of acid-catalysed ester hydrolysis of coal samples.*

9.2. Materials and methods

9.2.1. Sample material

For this study, three samples from the NZ coal mines and one from the DEBITS-1 well were chosen, representing low to moderate thermal maturation. The coal mine samples were: G001980 (Waikato Basin, R_0 0.44%), G001996 (West Coast, R_0 0.52%) and G001989 (Taranaki Basin, R_0 0.69%) (Tables 3.1 and 3.2), being in the early to main catagenesis maturity stage, the DEBITS-1 well sample was G004541 (Waikato Basin, R_0 0.29%) being in the stage of diagenesis (Table 3.3).

9.2.2. Experimental approach

The acidic hydrolysis was performed according to the procedure described in Chapter 4.1.4. To evaluate the temperature dependance of the hydrolysis four different reaction temperatures were chosen at 45°C, 60°C, 75°C and 90°C. For each coal, five samples were prepared for each temperature (in total 20 samples per coal), which were removed from the oven at different time periods after one, two, three, four and seven days of heating, respectively (Figure 9.1).

9. Kinetics Of The Hydrolysis Of Low Molecular Weight Organic Acids

Ion chromatography (IC) was performed to determine the amounts of hydrolytically released formate and acetate in the temperature series at the different time periods. For the calculation of the kinetic parameters, the concentration of the esters remaining in the kerogen matrix was used. To obtain the amount of the remaining esters, the previously determined total amounts of formate and acetate obtained by alkaline ester cleavage reaction were chosen (*cf.* Chapter 5, Table 5.1). The alkaline ester cleavage amounts were assumed to be the maximum ester concentration being releasable during maturation from the kerogen as discussed in Chapter 5. The amounts of remaining esters after each hydrolysis time step at a specific temperature were calculated by substracting the amounts of hydrolysed acids from the maximum ester concentrations.

9.2.3. Kinetic of acid ester hydrolysis

The hydrolysis of ester molecules is described in general by the following equation.

$$Ester + Water \rightleftharpoons Acid + Alcohol \qquad (9.1)$$

The esterification (formation of ester from acid and alcohol) is also catalyzed by protons, thus, the acid catalyzed ester hydrolysis is a reversible reaction. To force this reaction to the right side, either at least one starting substance has to be used in huge excess or at least one of the final products has to be removed from the reaction mixture. Such a removal can be achieved by pore water flows or active consumption of potential substrate by microorganisms.

The reaction speed (ν) depends on the reactant concentration (ester and water) and a reaction rate constant k (Equation 9.2). ν is also given by the change of the ester concentration with time t (Equation 9.3).

9.2. Materials and methods

$$\nu = k[ester][H_2O] \tag{9.2}$$

$$\nu = -\frac{d[ester]}{dt} \tag{9.3}$$

If water is used in huge excess, the change of the water concentration during the reaction is negligible and, therefore, the concentration of water at each time t ($[H_2O]_t$) equals the initial water concentration at $t=0$ ($[H_2O]_0$). Thus, the reaction rate only depends on the loss of ester (Equation 9.4) and with a new k', a so-called pseudo first order kinetic can be used. The combination of equation 9.2 and 9.3 considering 9.4 yields into equation 9.5.

$$k[ester][H_2O] = k'[ester] \tag{9.4}$$

$$-\frac{d[ester]}{dt} = k'[ester] \tag{9.5}$$

Transformation of equation 9.5 leads to equation 9.6. Integration in the limits between 0 (start of reaction) and t (any time during the reaction) and ester concentration at the start of the reaction ($[ester]_0$) and at any time of the reaction ($[ester]_t$) gives equation 9.8 and delogarithmized equation 9.9.

$$\frac{d[ester]}{[ester]} = -k'dt \tag{9.6}$$

$$\int_{ester_0}^{ester_t} \frac{d[ester]}{[ester]} = -k'\int_0^t dt \tag{9.7}$$

$$\ln \frac{[ester]}{[ester]_0} = -k't \tag{9.8}$$

$$[ester] = [ester]_0 e^{-k't} \tag{9.9}$$

9. Kinetics Of The Hydrolysis Of Low Molecular Weight Organic Acids

A plot of $\ln([ester]/[ester]_0)$ for the y-axis and the reaction time t for the x-axis produces a straight line with the negative reaction rate constant $-k'$ as slope, shown e.g. in Figure 9.2.

The hydrolysis reaction is a proton catalysed reaction. The concentration of the catalyst $[H^+]$ does not change during the reaction and can, thus, be considered as constant. Therefore, the reaction rate constant k' determined during the experiments is the constant for the given $[H^+]$ concentration or in other words the reaction rate constant at a given pH. Thus, the reaction speed ν (Equation 9.10) determined during the experiments also depends on the pH.

$$\nu = k'[ester] \tag{9.10}$$

$$\nu = k_0[ester][H^+] \tag{9.11}$$

$$k' = k_0[H^+] \tag{9.12}$$

The reaction rate constant k_0 is the part of k' deriving from the proton concentration provided by the autoprotolysis of the solvent water. Considering this, equation 9.10 gives equation 9.11 and, thus, k' is the product of k_0 and the additional proton from the catalyst (acid). The experiments were carried out with diluted HCl at pH 3 and, thus, the experimentally determined reaction rate constant k' is the constant for the hydrolysis at pH 3. In natural environments the groundwater pH is usually less acidic, but using equation 9.12, a k' for different pH values can be calculated by knowing the H^+ concentration.

The reaction constant k can be used to determine the half life ($t_{1/2}$) for formic ester or acetic ester using equation 9.13. For a first order reaction, $t_{1/2}$ is independent of the starting reactant concentration.

$$t_{1/2} = \frac{\ln 2}{k} \tag{9.13}$$

9.3. Results and discussion

9.3.1. Reaction rate constants of the acid hydrolysis to cleave formate and acetate from coal samples

In Figures 9.2 to 9.8, the amounts of remaining kerogen linked formic (A) and acetic acids (B) are shown at the different hydrolysis temperatures. The plots of $\ln([ester]/[ester_0])$ versus hydrolysis time t for formic acid (C) and acetic acid (D) were used to calculate the reaction rate constant k' for the hydrolysis at the given temperature and pH 3 from the slope of the linear regression (Equation 9.8). From this value, the reaction rate constants k' and half lives at pH 7 (being the mean pH value measured from the pore water of the DEBITS-1 well coals; Vieth et al., 2008) were calculated using Equations 9.12 and 9.13.

Tables A.1 to A.16 (Appendix) show the amounts of hydrolysed formic and acetic acids and the remaining formic and acetic acid esters within the coals after hydrolysis at 90°C (363.15 K), 75°C (348.15 K), 60°C (333.15 K) and 45°C (318.15 K) for 1, 2, 3, 4 and 7 days, respectively. In general, the residual kerogen-bound ester concentration decrease with increasing reaction time (*cf.* Tables A.1-A.4 and Figure 9.2).

Hydrolysis at 90°C

Results from the determination of $k'(pH\ 3)$ from linear regression for the hydrolysis of formic (Figure 9.2C) and acetic acid esters (Figure 9.2D) as well as the calculated constant k' at pH 7 are given in Table 9.1. For formate, samples G004541 and G001980 being in the diagenetic stage show generally higher experimentally determined k' at pH 3. Thus, the calculated half lives at pH 3 and 7 are significantly higher in the coals of the catagenetic maturity stage (Figure 9.3). This indicates that the cleavage of formic acid esters in the more mature samples is slower than in the immature samples. A

9. Kinetics Of The Hydrolysis Of Low Molecular Weight Organic Acids

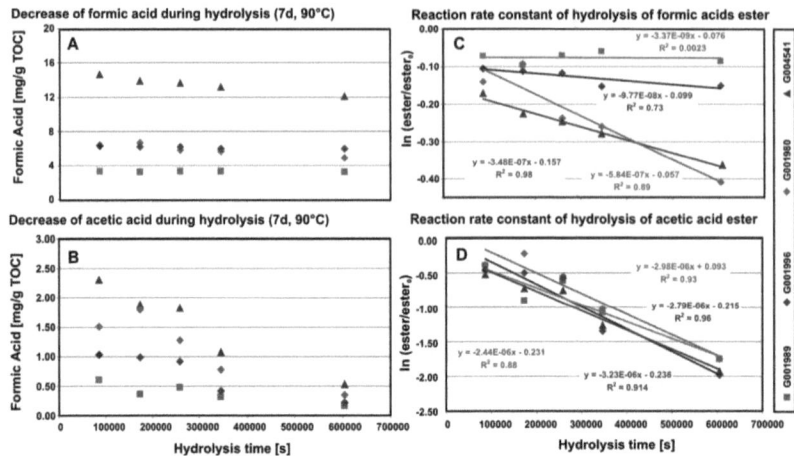

Figure 9.2.: *Residual kerogen-bound formic (A) and acetic acids (B) in the NZ coal samples during 7 d of hydrolysis and determination of k' for hydrolysis of formic (C) and acetic acid esters (D) ($90°C$, pH 3) by linear regression.*

reason for this might be that the ester bonds in the coals of higher maturity being already the remains of a long maturation history are sterically more protected than those in the immature sample material. As a consequence, the cleavage reaction is somewhat slower. The half life ($t_{1/2}$) of the ester concentration for the acidic hyrolysis at 90°C increases from the immature to the more mature samples (Table 9.1) also reflecting a higher stability of

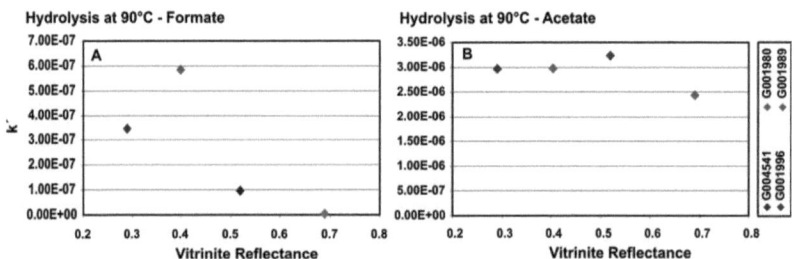

Figure 9.3.: *Reaction rate constants k' for hydrolysis of formic (A) and acetic esters (B) from the kerogen matrix of NZ coal samples of different maturity at $90°C$ and pH 3.*

the ester bonds in the mature samples most likely due to steric protection within the kerogen matrix.

Sample	G004541	G001980	G001996	G001989
R_0	0.29%	0.40%	0.52%	0.69%
Formate				
k' (pH 3)	$3.48 \cdot 10^{-07}$	$5.841 \cdot 10^{-07}$	$9.77 \cdot 10^{-08}$	$3.37 \cdot 10^{-09}$
$t_{1/2}$ (pH 3)	23.05 d	13.74 d	82.11 d	2380.57 d
k' (pH 7)	$3.48 \cdot 10^{-11}$	$5.84 \cdot 10^{-11}$	$9.77 \cdot 10^{-12}$	$3.37 \cdot 10^{-13}$
$t_{1/2}$ (pH 7)	631.60 a	376.36 a	2249.70 a	65221.23 a
Acetate				
k' (pH 3)	$2.97 \cdot 10^{-06}$	$2.89 \cdot 10^{-06}$	$3.23 \cdot 10^{-06}$	$2.44 \cdot 10^{-06}$
$t_{1/2}$ (pH 3)	2.70 d	2.69 d	2.48 d	3.29 d
k' (pH 7)	$2.97 \cdot 10^{-10}$	$2.89 \cdot 10^{-10}$	$3.23 \cdot 10^{-10}$	$2.44 \cdot 10^{-10}$
$t_{1/2}$ (pH 7)	74.01 a	73.76 a	68.05 a	90.08 a

Table 9.1.: *Determined reaction constant k' (pH 3) from the slope of the linear regression in figure 9.2, calculated k' (pH 7) using equation 9.8 and half life ($t_{1/2}$) using equation 9.13 of the ester concentration for the hydrolysis of formic and acetic acid ester in NZ coal samples at 90° C. R_0 = vitrinite reflectance.*

This trend was also slightly observed (although reaction rate constant k' of sample G001996 falls a little out of the trend) for the acetic acid esters (Figure 9.3). Generally, higher k' values indicate a higher reaction rate for the cleavage of acetate than for formate (Table 9.1). Additionally, the k' values of the immature and mature samples are not very different ((Figure 9.3b)) also leading only to small differences in the half life time (Table 9.1). The half lives for the ester cleavage reaction at pH 7 were calculated between 23.05 a (G004541, R_0 0.29%) and 2380.57 a (G001989, R_0 0,69%). These values appear to be very low regarding the suggested LMWOA release from the coals over geological time spans of millions of years presented in Chapter 5. Possible reasons for this are discussed below (Chapter 9.3.4). The calculated half lives for the cleavage reaction at the chosen laboratory conditions (pH 3) range from 2.70 d (G004541) to 3.29 d (G001989) demonstrating the

9. Kinetics Of The Hydrolysis Of Low Molecular Weight Organic Acids

acceleration of the hydrolysis reaction by the inceased proton concentration.

Hydrolysis at 75°C

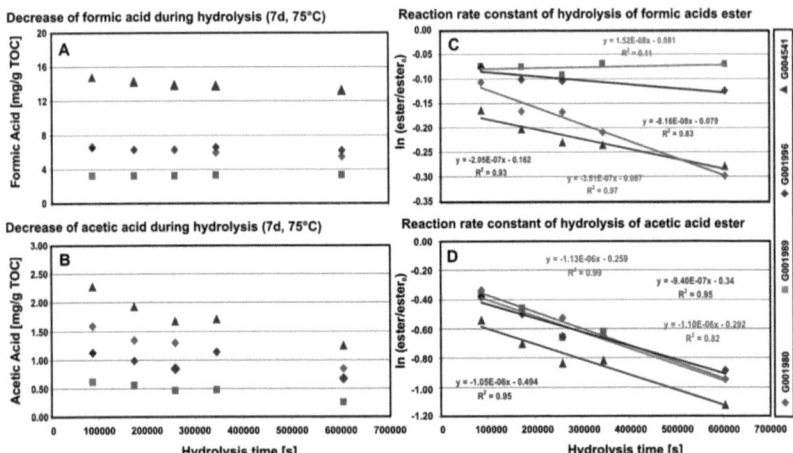

Figure 9.4.: Residual kerogen-bound formic (A) and acetic acids (B) in the NZ coal samples during 7 d of hydrolysis and determination of k' for hydrolysis of formic (C) and acetic acid esters (D) (75°C, pH 3) by linear regression. The value for sample G001989 at 7d appears to be unrealistic and are, therefore, not considered for the regression line.

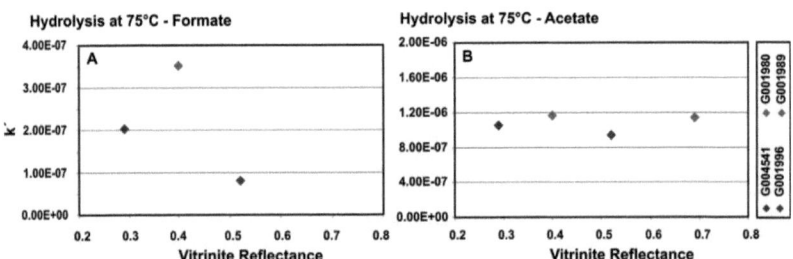

Figure 9.5.: Residual kerogen-bound formic (A) and acetic acids (B) in the NZ coal samples during 7 d of hydrolysis and determination of k' for hydrolysis of formic (C) and acetic acid esters (D) (75°C, pH 3) by linear regression.

9.3. Results and discussion

During the hydrolysis at 75°C for 7 days, the kerogen-bound esters remaining in the sediment show a slight decrease (cf. Tables A.5-A.8 and Figure 9.4). For the most mature sample G001989, containing the lowest amounts of ester-bound formate and acetate, no clear decrease could be observed regarding formate and, thus, for the formic acid ester hydrolysis, no valid k' and, therefore, no half life could be determined. For this sample, the straight line from linear regression gives an unrealistic positive slope value and, therefore, a negative constant k' (values in table 9.2 in brackets). However, at least this shows that at this increased level of maturity, the cleavage reaction is very slow, being within the error of the applied procedure. Half lives calculated for pH 7 range from 626.20 a (G001980) to 2693.57 a (G001996) for formate and 194.51 a (G001980) to 233.83 a (G001996) for acetate, which is again much lower as suggested for release of these acids from the coals during maturation over geological time spans (cf. Chapter 5).

Sample	G004541	G001980	G001996	G001989
R_0	0.29%	0.40%	0.52%	0.69%
Formate				
k' (pH 3)	$2.05 \cdot 10^{-07}$	$3.51 \cdot 10^{-07}$	$8.16 \cdot 10^{-08}$	$(-1.52 \cdot 10^{-08})$
$t_{1/2}$ (pH 3)	39.13 d	22.86 d	98.32 d	-
k' (pH 7)	$2.05 \cdot 10^{-11}$	$3.51 \cdot 10^{-11}$	$8.16 \cdot 10^{-12}$	$(-1.52 \cdot 10^{-12})$
$t_{1/2}$ (pH 7)	1072.17 a	626.20 a	2693.57 a	-
Acetate				
k' (pH 3)	$1.05 \cdot 10^{-06}$	$1.13 \cdot 10^{-06}$	$9.40 \cdot 10^{-07}$	$1.10 \cdot 10^{-06}$
$t_{1/2}$ (pH 3)	7.64 d	7.10 d	8.53 d	7.29 d
k' (pH 7)	$1.05 \cdot 10^{-10}$	$1.13 \cdot 10^{-10}$	$9.40 \cdot 10^{-11}$	$1.10 \cdot 10^{-10}$
$t_{1/2}$ (pH 7)	209.33 a	194.51 a	233.83 a	199.81 a

Table 9.2.: *Determined reaction constant k' (pH 3) from the slope of the linear regression in figure 9.2, calculated k' (pH 7) using equation 9.8 and half life ($t_{1/2}$) using equation 9.13 of the ester concentration for the hydrolysis of formic and acetic acid ester in NZ coal samples at 75°C. R_0 = vitrinite reflectance.*

Regarding the calculated k' at 75°C and pH 3, generally, the lower ma-

9. Kinetics Of The Hydrolysis Of Low Molecular Weight Organic Acids

ture samples show higher k'-values than the more mature samples. Thus, the calculated half life (Table 9.2) for the hydrolysis at pH 7 significantly higher at G001996. The calculated half lives for the hydrolysis of formate at 75°C and pH 7 are longer than those calculated for the hydrolysis at 90°C which can also be expected due to the lower energy supply. For acetate again only a small decrease of k' between the samples of lower and higher maturity can be observed (Figure 9.5). The resulting half life times are as for formate a little bit longer than in the 90°C experiments.

Hydrolysis at 60°C

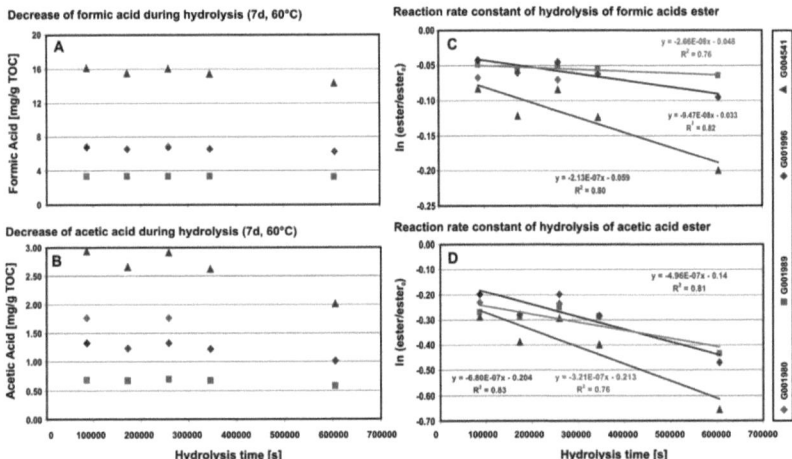

Figure 9.6.: *Residual kerogen-bound formic (A) and acetic acids (B) in the NZ coal samples during 7 d of hydrolysis and determination of k' for hydrolysis of formic (C) and acetic acid esters (D) (60° C, pH 3) by linear regression. (For the G001980 sample series only two values could be determined due to some experimental problems.)*

For the sample G001980 only two values were available (1 d and 3 d of hydrolysis time) due to some problems during sample preparation and measurement. Thus, for this sample the linear regression and the calculated half life are not utilisable. The kerogen-bound formate and acetate in the sediment slowly decrease in all samples with ongoing hydrolysis time (cf. Tables A.9-A.12 and Figure 9.6). The determined constant k' decreases with increasing maturation of the coal samples, again reflecting a slower hydrolysis reaction for the more mature samples (Table 9.3 and Figure 9.7). As a consequence, the calculated half lifes for the hydrolysis increase also with increasing maturity. The values are again higher than those calculated for 90°C and 75°C but ranging from 1031.90 a (G004541) to 8262.99 a (G001989) for formate and 323.23 a (G004541) to 684.72 a (G001989) they are still very low compared with the suggested LMWOA release during maturation (cf. Chapter 5).

Sample	G004541	G001980	G001996	G001989
R_0	0.29%	0.40%	0.52%	0.69%
Formate				
k' (pH 3)	$2.13 \cdot 10^{-07}$	-	$9.47 \cdot 10^{-08}$	$2.66 \cdot 10^{-08}$
$t_{1/2}$ (pH 3)	37.66 d	-	84.72 d	301.60 d
k' (pH 7)	$2.13 \cdot 10^{-11}$	-	$9.47 \cdot 10^{-12}$	$2.66 \cdot 10^{-12}$
$t_{1/2}$ (pH 7)	1031.90 a	-	2320.97 a	8262.99 a
Acetate				
k' (pH 3)	$6.80 \cdot 10^{-07}$	-	$4.96 \cdot 10^{-07}$	$3.21 \cdot 10^{-07}$
$t_{1/2}$ (pH 3)	11.80 d	-	16.17 d	24.99 d
k' (pH 7)	$6.80 \cdot 10^{-11}$	-	$4.96 \cdot 10^{-11}$	$3.21 \cdot 10^{-11}$
$t_{1/2}$ (pH 7)	323.23 a	-	443.14 a	684.72 a

Table 9.3.: *Determined reaction constant k' (pH 3) from the slope of the linear regression in figure 9.2, calculated k' (pH 7) using equation 9.8 and half life ($t_{1/2}$) using equation 9.13 of the ester concentration for the hydrolysis of formic and acetic acid ester in NZ coal samples at 60° C. R_0 = vitrinite reflectance.*

9. Kinetics Of The Hydrolysis Of Low Molecular Weight Organic Acids

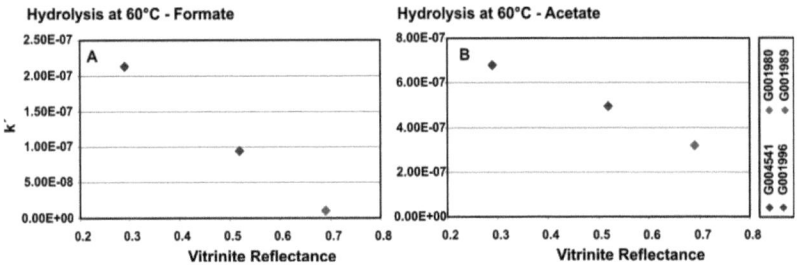

Figure 9.7.: *Residual kerogen-bound formic (A) and acetic acids (B) in the NZ coal samples during 7 d of hydrolysis and determination of k' for hydrolysis of formic (C) and acetic acid esters (D) (60° C, pH 3) by linear regression. (k' for G001980 could not be determined due to some experimental problems see figure 9.6)*

Hydrolysis at 45°C

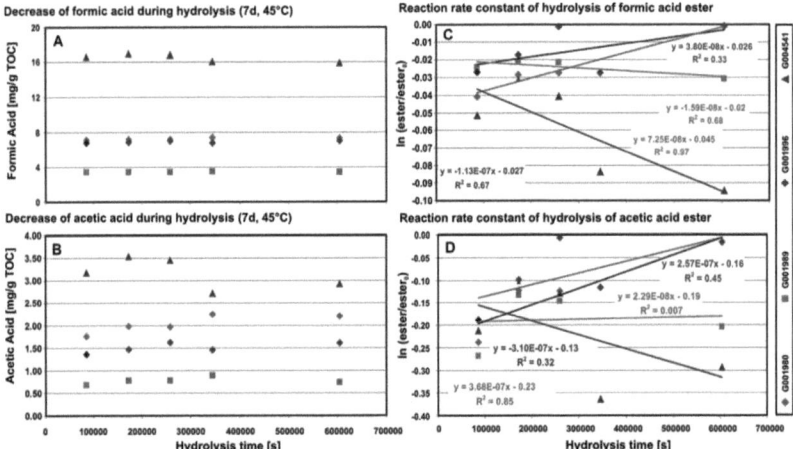

Figure 9.8.: *Residual kerogen-bound formic (A) and acetic acids (B) in the NZ coal samples during 7 d of hydrolysis and determination of k' for hydrolysis of formic (C) and acetic acid esters (D) (45° C, pH 3) by linear regression.*

Evalation of the hydrolysis results at 45°C is problematic. The amounts of cleaved formate and acetate during 7 days of hydrolysis are very low and high scattering of the determined values occurred, most likely due to variations in the experimental procedure and IC measurement. At this temperature in most cases no significant decrease of the ester-bound LMWOAs could be observed during 7 days. Therefore, the linear regression in Figure 9.8C and 9.8D produced often invalid values for k' (negative k' values) and half lifes could not be calculated (Table 9.4, invalid values for k' in brackets).

Sample	G004541	G001980	G001996	G001989
R_0	0.29%	0.40%	0.52%	0.69%
Formate				
k' (pH 3)	$1.13 \cdot 10^{-07}$	$(-7.25 \cdot 10^{-08})$	$(-3.80 \cdot 10^{-08})$	$1.59 \cdot 10^{-08}$
$t_{1/2}$ (pH 3)	71.00 d	-	-	504.56 d
k' (pH 7)	$1.13 \cdot 10^{-11}$	$(-7.25 \cdot 10^{-12})$	$(-3.80 \cdot 10^{-12})$	$1.59 \cdot 10^{-12}$
$t_{1/2}$ (pH 7)	1945.09 a	-	-	13823.62 a
Acetate				
k' (pH 3)	$3.00 \cdot 10^{-07}$	$(-3.68 \cdot 10^{-07})$	$(-2.57 \cdot 10^{-07})$	$(-2.29 \cdot 10^{-08})$
$t_{1/2}$ (pH 3)	26.74 d	-	-	-
k' (pH 7)	$3.00 \cdot 10^{-11}$	$(-3.68 \cdot 10^{-11})$	$(-2.57 \cdot 10^{-11})$	$(-2.29 \cdot 10^{-12})$
$t_{1/2}$ (pH 7)	732.65 a	-	-	-

Table 9.4.: *Determined reaction constant k' (pH 3) from the slope of the linear regression in figure 9.2, calculated k' (pH 7) using equation 9.8 and half life ($t_{1/2}$) using equation 9.13 of the ester concentration for the hydrolysis of formic and acetic acid ester in NZ coal samples at 45°C. R_0 = vitrinite reflectance.*

9.3.2. Critical evaluation of the determination of k'

The determination of k' showed best results when the amounts of cleaved LMWOAs are high. This was the case for higher temperatures when the cleavage reaction speed is highest. At 90°C, k' could be determined for all samples. When the amounts of cleaved LMWOAs decrease due to slower reaction rate and amounts of LMWOA in the initial sample material (low

9. Kinetics Of The Hydrolysis Of Low Molecular Weight Organic Acids

mature versus higher mature samples), the scattering of the measured concentrations becomes more significant, sometimes leading to invalid slope values from the linear regression of the ln([ester]/[ester]$_0$) versus reaction time plots. Therefore, for example at 45°C only for the most immature sample G004541, a valid k' could be calculated for formate and acetate.

Another problem of the applied methodological approach might be that the concentration of the LMWOAs ([ester]$_0$) at t=0 was determined using the alkaline ester cleavage reaction assumed to be more complete than the acidic hydrolysis. However, both types of cleavage reaction use a different reaction mechanism including sterically different transition states (Day and Ingold, 1941). Thus, the two different cleavage reactions may not be completely comparable and, therefore, the amount of formate and acetate at t=0 was not considered in the regression line to calculate k'.

Furthermore, it was recognised that it could be a problem that for each sample and temperature, only five experiments were carried out. Thus, only a maximum of five data points for each sample and temperature were available and scattering in the measured concentrations due to the experimental procedure and IC measurement (especially at low LMWOA concentrations) has a strong influence on the straight line from linear regression. Additionally, the time range could have been more extended, especially, for the reaction experiments at lower temperatures.

To overcome these problems, a significantly greater number of experiments should be carried out over a broader time range. The lowest temperature used here (45°C) appeared to be too low for a useful analysis at least at the time range chosen. Thus, in future experiments, the time should be extended and/or additional experiments should be carried out at higher temperatures. Due to the use of closed vials, temperatures up to 120°C should be possible. Furthermore, the implementation of replicate experiments would help to identify and eliminate outlier. However, this would extend the number of experiments from the current 80 to a much higher number, which could not be accomplished during this thesis due to the high demand of time needed.

However, although the exact values for k' and half life might be questionable due to experimental problems, the data show some interesting trends allowing a valuable insight into the ester cleavage processes in kerogen of different maturity. For formate, the reaction constant k' seems to be higher in the low mature samples than in the samples of higher maturity (Figures 9.3A, 9.5A and 9.7A) indicating a faster cleavage reaction process in the immature material. This points to a higher steric protection of the remaining ester bonds in the higher mature samples, reducing the reaction rate in these samples. For acetate, the difference of the ester cleavage between lower mature and higher mature samples appear to be smaller. Similar observations can be made for the half life of the kerogen-bound LMWOAs being dependent on the reaction constant k'. For formate and acetate, generally, the half life times increase for the higher mature samples, resembling the different ester bond stabilities. Concomitantly, this coincides with the observed rapid decrease of the kerogen-bound LMWOAs during diagenesis, slowing down during early catagenesis as shown in Chapter 5 Figure 5.3. Another aspect is that, generally, the reaction rate of the ester cleavage is faster at higher temperatures (especially for acetate), which is explained by the higher energy supply.

9.3.3. Temperature dependence of the reaction rate constant for the acid hydrolysis of formate and acetate

The reaction rate constant depends on the reaction temperature. A successful reaction takes place when the energy of the collision of the reactants is high enough to overcome the energy barrier resulting from the nucleus-nucleus repulsion. Higher temperatures lead to more energetic motion and to more collisions with higher kinetic energy causing into a greater chance to overcome the energy barrier of the reaction. Thus, higher temperatures lead to more successful collisions and the reaction is accelerated, resulting

9. Kinetics Of The Hydrolysis Of Low Molecular Weight Organic Acids

in a higher reaction rate constant. The temperature dependence of the reaction rate constant is described by the equation of *Arrhenius* (Equation 9.14) which is found by empirical observations.

$$k = Ae^{-\frac{E_A}{RT}} \quad (9.14)$$

E_A is defined as the activation energy. The factor A is called frequency factor and describes in the broadest sense the fractional amount of collisions that provides enough kinetic energy and leads to a successful reaction (Atkins, 1986) and R is the universal gas constant (being the product of the *Avogadro* constant (N_A) and the *Boltzmann* constant (k_B); 8.314 J mol^{-1}K^{-1}). Applying the logarithm to equation 9.14, the *Arrhenius*, equation can be written in the following way:

$$lnk = lnA - \frac{E_A}{R} \cdot \frac{1}{T} \quad (9.15)$$

Plotting $ln\ k$ against $1/T$ will give a straight line with the term $-E_A/R$ as slope and $ln\ A$ as intersect. Thus, E_A and A can be determined by linear regression. For the determination of E_A, the temperature has to be provided in K (Kelvin).

In Figure 9.9, the *Arrhenius* plots for the hydrolysis of formic acid ester and acetic acid ester are shown. The data used for these plots are given in Tables A.17 - A.20. The calculation of the activation energies (E_A) and frequency factors (A) are shown in Table 9.6.

In Table 9.5, all reaction rate constant k' values determined in Chapter 9.3.1 are listed. For formate, only for the immature sample G004541, k' values could be determined for all four temperatures (90°C, 75°C, 60°C and 45°C) applied in the kinetic experiments. For G001980, only two k' values are available, for G001996, the 60°C k' value is significantly higher than

k' formate pH 3				
Sample	G004541	G001980	G001996	G001989
R_0	0.29%	0.40%	0.52%	0.69%
90°C	$3.48 \cdot 10^{-7}$	$5.84 \cdot 10^{-7}$	$9.77 \cdot 10^{-8}$	$3.37 \cdot 10^{-9}$
75°C	$2.05 \cdot 10^{-7}$	$3.51 \cdot 10^{-7}$	$8.16 \cdot 10^{-8}$	-
60°C	$2.13 \cdot 10^{-7}$	-	$9.47 \cdot 10^{-8}$	$2.66 \cdot 10^{-8}$
45°C	$1.13 \cdot 10^{-7}$	-	-	$1.59 \cdot 10^{-8}$
k' acetate pH 3				
90°C	$2.97 \cdot 10^{-6}$	$2.89 \cdot 10^{-6}$	$3.23 \cdot 10^{-6}$	$2.44 \cdot 10^{-6}$
75°C	$1.05 \cdot 10^{-6}$	$1.13 \cdot 10^{-6}$	$9.40 \cdot 10^{-7}$	$1.10 \cdot 10^{-6}$
60°C	$6.80 \cdot 10^{-7}$	-	$4.96 \cdot 10^{-7}$	$3.21 \cdot 10^{-7}$
45°C	$1.13 \cdot 10^{-7}$	$3.00 \cdot 10^{-7}$	-	-

Table 9.5.: *Reaction rate constant k' values determined during the kinetic experiments. R_0 = vitrinite reflectance.*

for 75°C and for G001989, the reaction rate constant is lowest at the 90°C experiment. Thus, a determinaton of E_A and A seems only reasonable for sample G004541.

The obtained results from the determination of the activation energy, as well as the frequency factor, can only be considered as a rough estimation. The main problem is the discussed quality of the determined k' values that reveal huge variations and show in some cases invalid (negative) values as shown in sections 9.3.2 and 9.3.1. Nevertheless, the data for the activation energy and frequency factors show some interesting trends for the hydrolysis process in coal samples of different maturity.

For acetate, the calculated values of E_A, in general, show an increase with increasing maturation of the coal samples (Table 9.6). This indicates that the ester cleavage in the more mature samples requires higher energies than the cleavage in the younger samples. This derives most likely from structural differences within the kerogen of different maturity. In the immature samples, a larger part of the ester bonds are easier to reach for the

9. Kinetics Of The Hydrolysis Of Low Molecular Weight Organic Acids

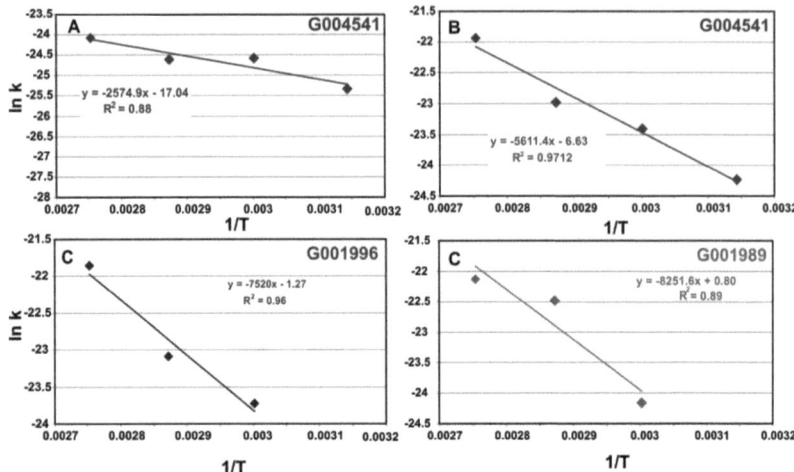

Figure 9.9.: *Arrhenius plot for the acidic hydrolysis of A) formic acid ester and B,C,D) acetic acid ester in NZ coal samples and linear regression to calculate the activation energy and frequency factor. Due to the lack of appropriate k', data activation energy and frequency factor could not be calculated for G001980 (formate and acetate), G001996 (formate) and G001989 (formate).*

hydrolysis processes in the pore water. Thus, the better reachable ester bonds are suggested to be hydrolysed first. During maturation, the structure of the kerogen becomes more and more compact. The remaining ester bonds within the coal samples of higher maturity are probably shielded by the increasingly compact macromolecular structure. If the frequency factor A is a measure for the number of collisions being needed for a successful reaction, it also can be seen that this number increases from the immature to the mature samples (Table 9.6).

9.3. Results and discussion

Sample	G004541	G001980	G001996	G001989
Formate				
E_A/R	2574.90	-	(106.65)	-
E_A [J/mol]	$2.14 \cdot 10^4$	-	$(8.87 \cdot 10^2)$	-
E_A [kcal/mol]	5.11	-	(0.21)	-
ln A	-17.041	-	(-25.10)	-
A	$3.97 \cdot 10^{-8}$	-	$(1.26 \cdot 10^{-11})$	-
Acetate				
E_A/R	5611.40	-	7520.00	8251.60
E_A [J/mol]	$4.67 \cdot 10^4$	-	$6.25 \cdot 10^4$	$6.86 \cdot 10^4$
E_A [kcal/mol]	11.14	-	14.93	16.38
ln A	-6.6264	-	-1.27	0.8042
A	$1.32 \cdot 10^{-3}$	-	0.28	2.23

Table 9.6.: *Calculation of E_A and A from linear regression (Figure 9.9) for the hydrolysis of formic and acetic acid ester in coal samples. 1 J = 4.1868 kcal, R = 8.314 J·mol⁻¹·K⁻¹.*

9.3.4. Problems in simulating the natural hydrolysis process in laboratory scale

The experiments to determine the kinetic parameters for the proton catalysed hydrolysis of formic acid and acetic acid esters within the kerogen were all performed using approximately 300 mg of dried and finely ground coal powder. The so prepared samples differ from the situation in the natural environment where the coals and lignites form massive layers. Within these layers, the permeability for the pore water is suggested to be significantly lower than for the powdered coals. However, including diffusion into the experiments would substantially extend the time needed for the experimental procedure.

The calculated half lifes of hydrolysis at pH 7 show values ranging from approximately 100 to several 10000 years. Although determined at different temperatures that were higher than those suggested to appear within sediments of early maturation stage, the calculated half lifes are significantly

9. Kinetics Of The Hydrolysis Of Low Molecular Weight Organic Acids

smaller than the observed release of LMWOAs from these lignite and coal samples over geological times of millions of years as suggested in chapter 5. Regarding the fact that LMWOAs were also found in samples of early and main catagenesis stage, it can be speculated that the release of LMWOAs in the natural environment is not only influenced by the kinetics of the hydrolysis reaction alone which would lead to a faster release than described in chapter 5. It can be speculated that additional processes may influence the release of LMWOAs into the pore water.

A reason for this could be the compaction of the sediments which may influence the accessibility of the esters and slow down the reaction rate. During the alteration of the sediment, the structure of the kerogen reorganises. As a result, some ester groups which were not available for cleavage reaction before may become accessible in older sediments, and others might be even more protected. They might be covered by hydrophobic layers of resins and hydrocabons or protected by wax and aliphatic biopolymers. Microenvironments might exist, where water is not in infinite supply. The reduced pore space within coaly layers could also inhibit the release of the ester cleavage products by pore water flow. The proton catalysed ester cleavage reaction is a reversible reaction and it may be speculated that remaining free LMWOAs in the pores, which were not transported off, may again form ester with free hydroxy groups and, therefore, the effective release of LMWOAs might be slowed.

Therefore, the chosen experimental procedure provides valuable insight into the stability trends of ester bonds in the kerogen structure of different maturity, but does not give information about the natural releasing process where pore space, permeability, pore water flow and diffusion might play a significant role. Nevertheless, the low E_A and A values point to an instant food release and changes are suggested to be triggered by the other parameters. Leaching experiments with intact coal samples might help to get more information about these influence factors. Furthermore, the direct influence of the deep biosphere has to be taken into account when simulating the natural release of these ester cleavage products, being a potential feedstock for deep microbial life.

9.4. Conclusion

In this chapter, the kinetic of the proton catalysed ester cleavage reaction in coal samples was investigated. Four coal samples of different thermal maturity (two from a diagenetic and two from an early to main catagenetic maturation stage) were examined applying acidic ester cleavage reaction at four different temperatures (90°C, 75°C, 60°C and 45°C) over a reaction time range of seven days.

Although the exact values for k', E_A and A might be questionable due to experimental problems, these investigations reveal interesting insights into the general trends of the ester cleavage reactions in organic matter of different maturity. The determined reaction rate constants k' for formate and acetate generally decrease with ongoing maturation of the samples, reflecting a slower cleavage reaction in the more mature samples. The determined activation energies (E_A) and the frequency factors (A) increase with ongoing maturation. This seems to reflect structural changes within the kerogen. It can be suggested that the ester bonds within the more mature samples are better sterically protected by the compact structure of the organic macromolecule and, therefore, higher activation energy and number of collisions is needed for the ester cleavage reaction.

Despite that, the hydrolysis experiments were carried out at comparatively high temperatures compared with the environmental conditions within the sediments, the calculated half lives ($t_{1/2}$) for the hydrolysis at pH 7 (ranging from approximately 100 to several 10000 years) appearing to be significantly too low to explain the observed release of LMWOAs from the NZ lignite and coal samples during millions of years as described in chapter 5. The observed process in the natural environment appears to be slower than deter-

9. Kinetics Of The Hydrolysis Of Low Molecular Weight Organic Acids

mined from the experiments in this study. Therefore, the proton catalyzed hydrolysis reaction alone can not be responsible for the observed timing of the release of LMWOAs. Some additional factors such as pore space, permeability, pore water flow and diffusion within the organic macromolecules as well as protection of ester groups by hydrophobic layers of aliphatic biopolymers, resins and hydrocarbons forming microenvironments where water is not in infinite supply are suggested to influence and slow down the abiotic release of LMWOAs within the natural environment. These factors should be considered in future experimental design to investigate the release of LMWOAs from the kerogen matrix at different maturity stages. Nevertheless, the observed increase of the calculated half lives with ongoing maturation also points to a slower LMWOA release in the more mature samples.

10. Ether Cleavage In New Zealand Coal Samples

10.1. Introduction

The maturation of sedimentary organic matter is characterised by the degradation of reactive functional groups within the organic matter. The most abundant hetero atom in sedimentary organic matter is oxygen, forming carboxyl (esters and acids), carbonyl (ketones and aldehydes), ether and hydroxyl functional groups. The compositional alteration of the organic matter with ongoing maturation is characterised by a loss of oxygen being indicated by a decrease of the O/C ratio of the organic matter as visualised in a van Krevelen diagram (van Krevelen, 1961).

To investigate changes of the structural composition within the macromolecular organic matter during maturation, selective chemical degradation procedures are appropriate tools used in organic geochemistry. A broad suit of chemical reactions is well established for a selective cleavage of different bond types (Rullkötter and Michaelis, 1990; Richnow et al., 1992; Schaeffer et al., 1995; Schaeffer-Reiss et al., 1998). Oxygen containing compound linkages within the kerogen matrix are ester or ether bonds. In a stepwise procedure, ester cleavage has to be performed prior to ether cleavage due to the fact that ether cleavage methods will also attack ester functions. Ester cleavage of a series of New Zealand lignites and sub-bituminous coals of different maturity (R_0 0.27 to 0.8%) from five various sedimentary basins in New Zealand (Northland, Eastern Southland, Waikato, West Coast and Taranaki Basin) (*cf.* Chapters 5 and 6) was already in the focus of two previous studies (Glombitza et al., 2009b,a).

In the current study, ether cleavage was applied to the same sample set to investigate the changes of ether-bound compounds with ongoing maturation. For the cleavage of ether bonds, several methods are available, such as the use of HI (Höld et al., 1998; Putschew et al., 1998; Schaeffer-Reiss et al., 1998), boron trichloride (BCl_3) (Chappe et al., 1982; Richnow et al., 1992), boron tribromide (BBr_3) (Schwarzbauer et al., 2003; Wollenweber et al., 2006) boron trifluoride (BF_3) (Narayanan and Lyer, 1965; Ambles et al., 1996), trifluoro acetic acid (Bhatt and Kulkarni, 1983) or trimethylsilyiodide (TMSI) (Michaelis and Richnow, 1989). For this study we have used the boron tribromide procedure, because the produced bromides are more stable than e.g. the iodides and can be analysed by GC-MS without a further reduction step.

10.2. Materials and Methods

10.2.1. Sample material

The sample material for the current study came from a set of coal samples gathered in 2002 from different coal mines in New Zealand. Several investigations of this coal sample series were previously reported. The kerogen composition determined from open system pyrolysis and Rock-Eval data were reported in Vu et al. (2008), low molecular weight organic acids from water extraction of these samples can be found in Vieth et al. (2008), the molecular composition of the aliphatic bitumen fraction obtained by solvent extraction is investigated in Vu et al. (2009), results of alkaline ester cleavage in the pre-extracted samples were reported in Glombitza et al. (2009b) (*cf.* Chapter 5) for the low molecular weight organic acids (LMWOAs) and in

Glombitza et al. (2009a) (*cf.* Chapter 6) for the high molecular weight organic acids (HMWOAs).

For this study we selected 12 lignite and coal samples of different maturity ranging between R_0 0.27% and 0.8%. Two samples from the Northland (G001986 and G001988) and two samples from the Eastern Southland Basin (G001976 and G001978) are in the lignite stage. Three samples from the Waikato Basin (G001980, G001982 and G001983), three samples from the West Coast Basin (G001993, G01995 and G001996) and two samples from the Taranaki Basin (G001991 and G001994) are sub-bituminous coals (*cf.* Tables 3.1 and 3.2).

10.2.2. Methods

Sample preparation

The dried and finely ground coal samples were extracted with water for one day using Soxhlet extraction, followed by a three day Soxhlet extraction using an azeotrope solvent mixture of acetone (38%), methanol (30%) and chloroform (32%), and an additional extraction with pure methanol for one day. Furthermore, alkaline ester cleavage reaction was performed (Glombitza et al., 2009b,a). The residue of this ester cleavage procedure was used for the BBr_3 ether cleavage that is described below.

Ether cleavage procedure

The ether cleavage reaction was applied as reported by Wollenweber et al. (2006) using BBr_3 as reagent. Approximately 150 mg of the coal sample was weighed into a 10 ml GC vial, equipped with an magnetic stirrer. 5 ml of a BBr_3 solution (1 M in DCM, *Alfa Aescar GmbH & Co KG*, Karlsruhe, Germany) were added and the vial was sealed. The suspension was than placed in an ultrasonic bath for 2 h, than stirred for 24 h at room temperature and finally ultrasonicated for another 2 h. Finally, the vial was opened

10. Ether Cleavage In New Zealand Coal Samples

and approximately 3 ml of diethyl ether were carefully added dropwise while stirred until the excess of BBr_3 was deactivated. The residue was filtered using Whatmann glass fiber filters (poresize 0.7 μm) and washed three times with 5 ml of diethyl ether. The combined organic phases were washed three times with water and dried with Na_2SO_4 and the solvent was reduced at ambient pressure to approximately 2 ml.

GC-MS

Prior to GC-MS analysis 5 μg of 5-α-androstane were added to each sample as injection standard for quantification. To determine the GC retention times of the mono-bromoalkanes, a standard mixture of 1-bromododecane, 1-bromotetradecane, 1-bromohexadecane and 1-bromooctadecane was analysed. GC-MS analysis was performed using a Thermo Trace GC Ultra (Thermo Fisher Scientific Inc., Waltham, MA, USA) coupled to a Thermo DSQ mass spectrometer. Oven temperature was set to 50°C (1 min isothermal), heated ar a rate of 3°C/min to 310°C and held isothermal for 30 min at 310°C. Samples were injected with a PTV splitess injector from 50°C to 300°C with 10°C/s. The column used was a BPX5, length 50 m, diameter 0.22 mm and 0.25 μm film thickness. Helium was used as carrier gas. Mass spectra were obtained in the electron impact mode at 70 eV. The applied scan interval was 50-650 amu with 2.5 scans/sec, temperature of the ion source was set to 230°C. To identify brominated alkanes, the mass ion chormatogram m/z = 135 was used (being the mass of the characteristic pentacyclic fragment $C_4H_8Br^+$).

10.3. Results and discussion

10.3.1. Qualitative evaluation of the BBr$_3$ cleavage products from the New Zealand coal series

The investigated extracts obtained from BBr$_3$ ether cleavage contained various amounts of previously ether bound compounds and some ester bound

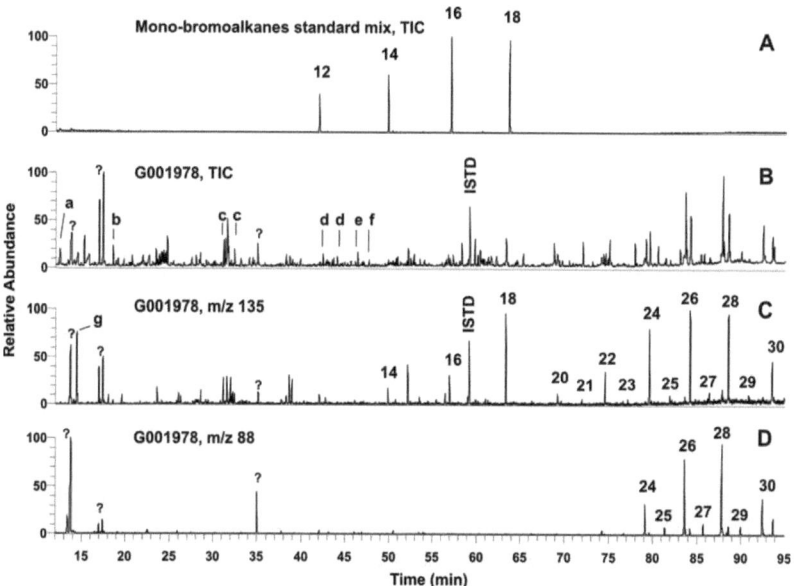

Figure 10.1.: *(A) Total ion current (TIC) chromatogram of a mono-bromoalkane standard mix (n-C$_{12}$, n-C$_{14}$, n-C$_{16}$, n-C$_{18}$) and (B) TIC-chromatogram of BBr$_3$ cleavage products from sample G001978 (Eastern Southland Basin, R$_0$ 0.28%), (C) all aliphatic brominated compounds with more than three carbon atoms (mass trace m/z 135), (D) ethyl esters (mass trace m/z 88). Numbers indicate the carbon atoms of the aliphatic chain; small letters indicate the folloing compounds, a: CH$_2$BrCOOC$_2$H$_5$, b: CO(CH$_2$Br)$_2$, c: C$_5$H$_9$Br$_3$, d: C$_6$H(CH$_3$)$_3$Br$_2$, e: C$_6$H$_2$Br$_3$OH, f: C$_6$H$_2$Br$_2$C(CH$_3$)$_3$OH, g: C$_4$H$_8$Br$_2$.*

10. Ether Cleavage In New Zealand Coal Samples

compounds that appeared to have survived the previously applied ester cleavage procedure (Tables 10.1 and 10.2). The ether cleavage products comprise several groups of compounds such as mono-bromoalkanes, poly-bromoalkanes, ethyl esters and brominated esters and ketones. In addition, some aromatic and phenolic compounds could be detected (Table 10.2). Figure 10.1 shows an example chromatogram (Figure 10.1B) of the products obtained from sample G001978 (Eastern Southland Basin). The mass trace m/z 135 was used to identify the brominated alkanes with aliphatic chains containing more than 3 carbon atoms (Figure 10.1C) and the mass trace m/z 88 was used to identify the ethyl esters (Figure 10.1D). For the identification of the aliphatic mono-bromoalkanes the chromatogram of a standard mix (containing n-C_{12}, n-C_{14}, n-C_{16} and n-C_{18} 1-bromoalkanes) is shown (Figure 10.1A).

Mono- and poly-brominated alkanes

Mono-bromoalkanes were found to show quite identical mass spectra. The signals of the different molecular ions were not detectable (Appendix, Figure B.1). Therefore, to identify the mono-bromoalkanes, a standard mixture of commercial available mono-bromoalkanes (n-C_{12}, n-C_{14}, n-C_{16} and n-C_{18}, Sigma-Aldrich Chemie GmbH, Buchs, Switzerland) was measured for comparison. Thus, the mono-bromoalkanes obtained from the lignite and coal samples were identified according to their GC retention times.

The released mono-bromoalkanes are cleavage products of formally ether- or ester-bound n-alcohols. They range from n-C_{14} to n-C_{30} with a strong predominance of even carbon numbers (Table 10.1). This coincides with the ether-cleavage product distributions found in other studies using different ether-cleavage procedures (Richnow et al., 1992; Höld et al., 1998; Schaeffer-Reiss et al., 1998). In other studies, applying the BBr$_3$ procedure the brominated alkanes were only described as bulk parameters due to their indistinguishable mass spectra (Wollenweber et al., 2006). In the Northland

10.3. Results and discussion

and Eastern Southland Basin samples, maxima in the distribution of the mono-bromoalkanes occur at n-C_{18} and n-C_{28}. The sub-bituminous coal samples from the Waikato, West Coast and Taranaki Basins contained lower numbers of mono-bromoalkanes. Only the n-C_{18} bromoalkane was found in almost all samples investigated (Table 10.1).

The mono-bromoalkanes were linked to the organic matter via one ether (or etser) bridge and, therefore, they represent terminal bound alcohols. These structural units are proposed to derive from plant waxes such as cutin and were also described by Wollenweber et al. (2006) after BBr_3 treatment of organic matter in Paleozoic coals of different locations in Europe, Asia, South America and Antarctica.

Additionally, several poly-brominated alkanes were found in almost all samples investigated. In contrast to the mono-brominated alkanes, these compounds have characteristic mass spectra (Appendix, Figures B.2 to B.9). The highest number of compounds was detected in the Eastern Southland lignite samples with tribromo-propane ($C_3H_5Br_3$, sample G001976 only), tetrabromo-propane ($C_3H_4Br_4$, sample G001976 only), dibromo-butane ($C_4H_8Br_2$, sample G001978 only), two regioisomers of tribromo-pentane ($C_5H_9Br_3$) and tetrabromo-pentane ($C_5H_8Br_4$, sample G001976 only) (Table 10.1). The two isomers of tribromo-pentane showed the highest concentrations of all poly-brominated alkanes. Only these two components were found to be present in all samples.

10. Ether Cleavage In New Zealand Coal Samples

Compound	G001986	G001988	G001978	G001976	G001982	G001983	G001980	G001996	G001995	G001994	G001993	G001991
R_0 [%]	0.27	0.27	0.28	0.29	0.40	0.41	0.44	0.52	0.52	0.61	0.76	0.80
Basin	NB	NB	ESB	ESB	WB	WB	WB	WCB	WCB	TB	WCB	TB
Mono-bromoalkanes												
C_{14}	58.80	48.99	17.32	15.10	-	-	-	-	-	-	-	-
C_{16}	10.45	15.00	9.14	9.25	-	-	-	-	5.09	6.48	5.77	-
C_{18}	48.56	49.87	32.25	29.30	-	22.79	28.70	11	13.15	10.88	7.13	7.93
C_{20}	32.73	25.94	13.52	13.01	-	-	-	-	-	-	-	-
C_{22}	35.96	26.73	12.63	16.23	-	-	-	-	-	-	-	-
C_{24}	37.95	37.88	35.67	47.57	-	-	-	-	-	-	-	-
C_{26}	81.57	52.52	71.67	40.00	-	-	-	-	-	-	-	-
C_{28}	131.37	106.52	70.26	43.73	-	-	-	-	-	-	-	-
C_{30}	58.29	33.89	48.83	42.46	-	-	-	-	-	-	-	-
Total	495.68	397.34	311.29	256.65	-	22.79	38.80	11	18.24	17.36	12.9	7.93
Di-,tri- and tetra-bromoalkanes												
$C_3H_5Br_2$	-	-	-	20.18	-	-	-	-	-	-	-	-
$C_3H_4Br_4$	-	-	-	14.20	-	-	-	-	-	-	-	-
$C_4H_8Br_2$	-	-	10.84	-	-	-	-	-	-	-	-	-
$C_5H_9Br_3$*	393.64	236.72	31.08	183.49	92.57	132.80	70.70	167.02	20.31	193.19	224.58	193.99
$C_5H_9Br_3$*	327.91	188.38	25.87	166.08	85.31	116.94	61.37	143.68	21.56	210.22	205.39	238.60
$C_5H_8Br_4$	-	-	-	5.43	-	-	6.62	-	-	-	-	-
Total	721.55	425.10	67.79	389.38	177.88	249.74	138.69	310.70	41.87	422.41	429.97	432.59

Table 10.1.: Monobromo-, dibromo-, tribromo- and tetrabromo-alkanes after BBr_3 ether cleavage experiments from NZ coal samples ranging from diagenesis to the main catagenesis (0.27 - 0.8% R_0) stage. Data are presented in µg/gTOC, R_0: Vitrinite Reflectance, NB: Northland Basin, ESB: Eastern Southland Basin, WB: Waikato Basin, WCB: West Coast Basin, TB: Taranaki Basin, *regioisomers or diastereoisomers

Compound	G001986	G001988	G001978	G001976	G001983	G001980	G001996	G001995	G001994	G001993
R_0 [%]	0.27	0.27	0.28	0.29	0.41	0.44	0.52	0.52	0.61	0.76
Basin	NB	NB	ESB	ESB	WB	WB	WCB	WCB	TB	WCB
Brominated esters and ketones										
$CO(CH_2Br)_2$	-	-	21.17	-	-	-	-	-	-	-
$CH_2BrCO_2C_2H_5$	843.25	761.92	36.98	30.34	15.41	13.05	-	-	-	-
$C_2H_4BrCH(CO_2CH_3)CH_3$*	198.01	260.05	-	15.32	-	-	-	-	-	-
$C_2H_4BrCH(CO_2CH_3)CH_3$*	268.53	480.05	-	-	-	-	-	-	-	-
Total	**1309.79**	**1502.02**	**58.15**	**45.66**	**15.41**	**13.05**	-	-	-	-
Aromatics										
$C_6H(CH_3)_3Br_2$*	-	-	12.27	12.34	-	5.41	6.76	3.37	8.48	9.43
$C_6H(CH_3)_3Br_2$*	-	-	10.11	7.80	-	5.85	5.72	5.11	9.19	8.73
Total	-	-	**22.38**	**20.14**	-	**11.26**	**12.48**	**8.48**	**17.66**	**18.16**
Phenols										
$C_6H_2Br_3OH$	-	-	15.87	-	-	-	-	-	-	-
$C_6H_3Br_2OH$	-	-	-	-	-	6.61	-	-	-	-
Total	-	-	**15.87**	-	-	**6.61**	-	-	-	-
Ethyl esters										
C_{24}	29.14	37.62	50.96	38.03	-	3.97	-	-	-	-
C_{26}	26.01	42.08	84.04	61.86	-	15.88	-	-	-	-
C_{28}	53.74	57.24	100.76	65.33	-	21.67	-	-	-	-
C_{30}	12.13	13.24	63.18	46.42	-	-	-	-	-	-
Total	**121.02**	**150.18**	**298.94**	**211.64**	-	**41.52**	-	-	-	-

Table 10.2.: Brominated esters and ketones, aromates, phenols and ethyl esters after BBr_3 ether cleavage experiments from NZ coal samples ranging from diagenesis to the main catagenesis (0.27 - 0.8% R_0) stage. Samples G001982 (R_0 0.4%) and G001991 (R_0 0.8%) are not mentioned due to the absence of these compounds. Data are presented in $\mu g/gTOC$, R_0: Vitrinite Reflectance, NB: Northland Basin, ESB: Eastern Southland Basin, WB: Waikato Basin, WCB: West Coast Basin, TB: Taranaki Basin, *regioisomers or diastereoisomers.

10. Ether Cleavage In New Zealand Coal Samples

Wollenweber et al. (2006) also described the finding of huge amounts of poly-brominated alkanes. Additionally, they identified di-bromoalkanes with chain length of 10 to 20 carbon atoms in some samples, which were not detected in the New Zealand coals investigated in this study. The polybrominated alkanes are suggested to form important cross-linking structures within the macromolecular organic matter by the formation of several ether bridges. Thus, they appear to be important units connecting aromatic substructures and forming the complex network structure. However, during the BBr_3 treatment, a possible transformation of hydroxy groups to bromine substituents has do be considered and, therefore, some of the bromine substituents might be former hydroxy groups (Wollenweber et al., 2006). In the ester fraction, obtained after saponification, additional cross-linking structures forming bridges in the macromolecular kerogen network are provided by dicarboxylic low and high molecular weight organic acids (as outlined in chapters 5 and 6).

Aromatics, phenols, ethyl esters and brominated esters and ketones

Two regioisomers of dibromo-trimethylbenzene ($C_6H(CH_3)_3Br_2$) were found in the Eastern Southland Basin lignites samples and in one of the Waikato Basin coals (G001980) as well as in the three West Coast Basin coals (G001993, G001995, G001996) and in one of the Taranaki Basin samples (G001994), but they were absent in all other samples investigated. The phenol 2,4,6-tribromo-phenol ($C_6H_2Br_3OH$) was only found in one of the Eastern Southland Basin lignites (G001978), dibromo-phenol ($C_6H_3Br_2OH$) was only found in sample G001980 from the Waikato Basin.

It might be speculated that the detected aromatic compounds are most likely related to altered substructures of lignin or suberin although they are not exactly matching the structural properties of lignin dominantly consisting of phenylpropyl units. Wollenweber et al. (2006) also reported the findings of several aromatics and phenols and correlated them to the maceral

10.3. Results and discussion

composition. The treatment of inertinite rich samples did not release these compounds or only in minor amounts. Some of these compounds were released from vitrinite rich samples but the highest amounts were obtained from liptinite rich samples. Because liptinite is not predominately derived from woody tissue they concluded that a primary lignin binding for these compounds cannot be assumed.

Nevertheless, these structural units appear to be fragments of the macromolecular network. The former linkage positions are indicated by the position of the bromine substituents. In figure 10.2, A selected aromatic substructures detected in the samples are shown.

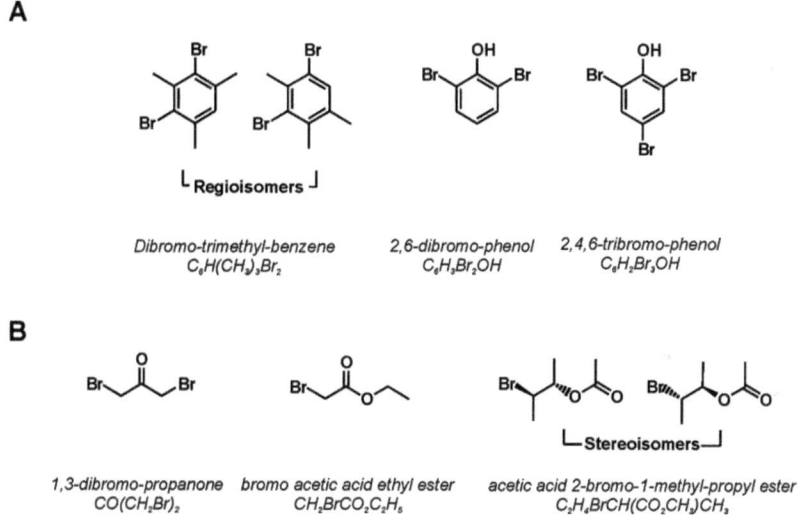

Figure 10.2.: *Selected structures of detected aromatics, phenols, ethyl esters and brominated esters and ketones from the New Zealand coal samples forming sub-structures within the macromolecular organic matter. The position of bromine reveals the former ether linkage, the ethyl ester is suggested to derive from former ester linkage.*

Ethyl esters with even carbon numbers in the range of n-C_{24} to n-C_{30} were found in the lignites from the Northland and Eastern Southland Basins as well as in sample G001980 from the Waikato Basin (with exception of n-

10. Ether Cleavage In New Zealand Coal Samples

C_{30}). They are related to terrestrial plant derived fatty acids, linked to the kerogen matrix via an ester bond. Although all samples were previously treated by alkaline ester cleavage reaction as described in Glombitza et al. (2009a) (*cf.* Chapter 6), these fatty acids obviously had remained within the macromolecular matrix. It can be speculated that the cleavage of ether-bound compounds and with that the breaking of smaller substructures within the kerogen exposes prior protected esters functionalities for cleavage reaction. The BBr_3 cleavage of these esters led to highly reactive acid bromides. The following addition of diethylether resulted finally in the formation of ethyl esters. This could explain why no other fatty acid esters than the ethyl esters were detected. Surprisingly, only long chain ethyl esters were detected. A reason why there are no short chain ethyl esters (n-C_{16}, n-C_{18}) is puzzling especially because the short chain fatty acids are the dominating fatty acids in the fatty acid distribution in the kerogen bound fraction (Figure 6.2). As discussed in Chapter 7 in the bitumen fraction, the short chain fatty acids are less dominant. However, if the ethyl ester represent bitumen fatty acids being released from some inner inclusions in the kerogen matrix after ether cleavage, at least a small proportion of short chain fatty acids can be expected. Thus, this observation might support the previously drawn suggestion that the short chain fatty acids of the kerogen matrix have a specific origin.

The ketone 1,3-dibromo-propanone ($CO(CH_2Br)_2$) was only detected in one of the Eastern Southland lignites samples (G001978). The bromo acetic acid ethyl ester ($CH_2BrCO_2C_2H_5$) was found in all samples of the Northland, Eastern Southland and Waikato Basins except in sample G001982 (Waikato Basin). This compound most likely derives from α-hydroxy acetic acid linked via an ester bond to the kerogen matrix, resulting in an ethyl ester during the BBr_3 cleavage procedure (see above), while the hydroxy group may have been linked via an ether bond to the kerogen. Furthermore, two stereoisomers of acetic acid 2-bromo-1-methyl-propyl ester ($C_2H_4BrCH(CO_2CH_3)CH_3$) were found in the lignites from the Northland Basin and in low amount in one of the Eastern Southland Basin samples (G001976). These compounds (except the acetic acid 2-bromo-1-methyl-propyl ester) are also most likely cross-linking structures within the kerogen matrix (the structures are shown in figure 10.2B).

The findings of poly-brominated alkanes, aromatics, phenols as well as brominated esters and ketones reveals a deep insight into the network structure of the organic macromolecules. Linkages of structural subunits in the kerogen by oxygen functionalities are characterised by various ether-bridges of short chain length bearing sometimes additional oxygen functionalities such as ketones. Thus, together with some low and high molecular weigth organic dicarboxylic acids, especially oxalic acid (cf. Chapter 5, Glombitza et al. 2009b), ether lipids form an important cross-linking bridge within the network.

10.3.2. Quantitative evaluation of the BBr_3 cleavage products from the New Zealand coals

Mono- and poly-brominated alkanes

The highest concentrations of mono-bromoalkanes were found in the Northland and Eastern Southland Basin samples. A strong decrease of these compounds is observed during early diagenesis. In the early and main catagenesis stage samples, only small concentrations are detectable (Figure 10.3A). Although facies differences have to be considered for the immature samples, this indicates a rapid degradation of terminal ether bound alcohols during diagenesis.

The maturation related decrease of the ether-bound moieties (Figure 10.3) is comparable to the observed decrease of the ester-bound alcohols (cf. Figure 6.3). This result is surprising because esters are thought to be less stable to thermal decomposition than ethers. The thermal decomposition of ethers proceeds via either a four center transition state or a free radical chain

10. Ether Cleavage In New Zealand Coal Samples

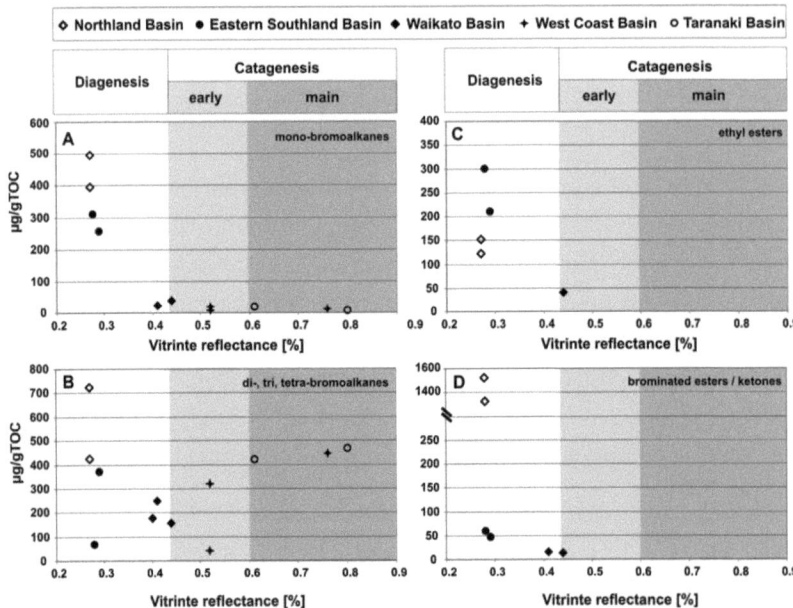

Figure 10.3.: *Products of BBr₃ ether cleavage of NZ lignite and coal samples of different maturity, grouped into mono-brominated alkanes (A), poly-brominated alkanes (B), ethyl esters (C) as well as brominated esters and ketones (D). Note different y-axis.*

decomposition which should require more energy than the decomposition of esters by a quasi six-membered ring transition state (Blake et al., 1961). Despite that, the observed differences in decomposition temperatures for various esters and ethers are small. Blake et al. (1961) measured ranges of approximately 270 to 350°C for esters and 200 to 450°C for ethers depending on different substituents. Therefore, it may be speculated that the difference in stability between ester and ether bonds, at least for the terminal bound compounds, is not strong enough to be observed in a maturation series representing alterations within geological time spans.

Interestingly, the poly-brominated alkanes do not show the same maturity related decrease as observed for the mono-bromoalkanes (Figure 10.3B).

10.3. Results and discussion

Highest amounts were also found in the Northland and the Eastern Southland Basin samples with decreasing concentrations during early diagenesis. With the exception of two samples (G001978 and G001995), showing extremely low values, a small minimum at still comparably high concentrations in the distribution of poly-brominated alkanes is observed during the late diagenesis to early catagenesis samples from the Waikato Basin. In contrast to the mono-bromoalkanes the poly-brominated alkanes occur in relatively high concentrations in the catagenesis samples from the West Coast and Taranaki Basins. Their distribution remains relatively constant or might even slightly increase. This indicates that after early diagenesis, when facies related differences play a more important role (*cf.* Chapter 6), the concentrations of these compounds are not decreasing with increasing maturation level (at least in the range investigated here). Due to their suggested function as linkage structures within the macromolecular network, it can be speculated that they are mainly located within the network concomitantly being sterically protected by the organic matrix and, therefore, are less effected by alteration processes than the terminal ether-bound aliphatic moieties.

In contrast to the ether bridges, the ester bridges of long chain dicarboxylic acids were only found in the early diagenetic stage samples of the Northland and the Eastern Southland Basins in significantly smaller amounts (*cf.* Chapter 6.7, Table 6.2) than the ether link-bridges described above. The short chain dicarboxylic acid (oxalic acid) also decreases rapidly during diagenesis and early catagenesis, but is still present in significant amounts at higher maturities (*cf.* Figure 5.3). This suggests that within the organic macromolecular matrix, link-bridges formed by short chain ether and ester bonds are considerably more important and stable than those formed by long chain compounds, which seem to decrease rapidly during early diagenesis.

10. Ether Cleavage In New Zealand Coal Samples

Aromatics, phenols, ethyl esters and brominated esters and ketones

Aromatics and phenols identified from the samples do not show any relationship with maturation (Table 10.2). They are probably internal aromatic substructures of the macromolecular organic matter and, therefore, they might be more facies dependent and less effected by maturation processes.

The ethyl esters, formed by long chain fatty acids, only show significant amounts in the diagenetic stage samples (Figure 10.3C) and they are absent in the catagenetic stage samples. Therefore, they show a different pattern with ongoing maturation as the kerogen bound fatty acids (Figure 6.5) and also if compared to the distribution of long chain fatty acids in the bitumen of the selected coal samples, distinct differences can be seen (Figure 7.3). Their amounts are in the same order of magnitude as the amounts of the kerogen-bound and the free bitumen FA's (cf. Tables 6.1, 7.1 and 10.2). This also suggests that the ethyl esters repesent a different fatty acid fraction which is not accessible by organic solvent extraction or by alkaline ester cleavage procedure. Together with the fact that no short chain fatty acid ethyl esters were detected (as discussed above), the origin of these ethyl esters remains unclear. They might represent an inner kerogen fraction being strongly dominated by a plant signature.

The highest amounts of bromo acetic acid ethyl ester ($CH_2BrCO_2C_2H_5$) were found in the lignites from the Northland Basin with approximately 20 times higher concentrations than in the Eastern Southland Basin samples. The Waikato Basin samples released approximately half of the concentrations obtained from the Eastern Southland Basin samples (Table 10.2). The huge difference between the Northland and Eastern Southland samples is most likely facies related rather than the result of thermal maturation because the difference in thermal maturation between the Northland (R_0 0.27%) and the Eastern Southland Basin samples (R_0 0.28/0.29%) is only small. On the other hand, the fact that this compound was absent in the West Coast and Taranaki Basin samples indicates that its degradation appears also during diagenesis to early catagenesis (Figure 10.3D). These findings combined with the fact that e.g. 1,3-dibromo-propanone ($C_3H_4OBr_2$) and acetic acid 2-bromo-1-methyl-propyl ester ($C_2H_4BrCH(CO_2CH_3)CH_3$) were only found in the Northland Basin samples in significant concentrations suggest that the linkages structures highly depend on the chemical constitution of the contributing source material and that these compounds are rapidly degraded during diagenesis.

10.4. Conclusion

The BBr_3 cleavage of kerogen ether bound compounds in lignite and sub-bituminous coal samples from the Northland, Eastern Southland, Waikato, West Coast and Taranaki Basins of New Zealand resulted in a complex set of different ether cleavage products such as mono- and poly-brominated alkanes, brominated aromatic compounds, ethyl esters and brominated esters and ketones.

Mono-brominated alkanes deriving from terminal ether-bound alcohols were found in strong decreasing concentrations related to the ongoing maturation of the samples. The poly-brominated alkanes were found in all samples and are suggested to represent important cross-link bridges within the kerogen. Brominated esters and ketones also represent cross-linking structures, but the observed distribution of these compounds suggests a higher dependence to the contributing plant facies and a higher sensitivity to maturation related degradation than the polybrominated alkanes. Polybrominated aromatic compounds appear to represent substructures or also bridges in the macromolecular network of the kerogen. The additionally detected ethyl esters might represent former ester bound acids that had survieved saponification and were transformed via acid bromides into ethyl esters during the ether cleavage procedure. Their different ditribution from kerogen-bound fatty acid fraction releasable by saponification as well as their relatively high amounts suggest that they represent an inner kerogen fatty acid fraction which is not accessible by the alkaline ester cleavage procedure.

The lignite samples of the Northland and Eastern Southland Basins showed significantly higher contents of mono-brominated alkanes than the sub-bituminous coal samples from the Waikato, West Coast and Taranaki Basins indicating a major loss of these terminal ether-bound alcohols during diagenesis. Although ether bonds are known to be more stable than ester bonds, the main decomposition of ether- and ester-bound alcohols can be observed in the same range during diagenesis. Thus, it is speculated that the stability differences between both linkages are not strong enough to be observed in alteration processes.

In contrast, the concentrations of the poly-brominated alkanes, representing cross-linking short chain alcohols, remain, after a small decrease during diagenesis, relatively high during the early and the beginning of the main catagenesis. This suggests that these compounds are important linkage structures within the kerogen being sterically shielded within the macromolecular network. Together with short chain ester moieties such as oxalic acid these short chain ether link-bridges connecting the substructures of the macromolecular network appear to be of higher relevance than the cross-linking long chain dicarboxylic acids which were only found in the early diagenesis stage samples after saponification.

11. Conclusions

Goals of this study

The discovery of widespread microbial life in the deep subsurface rises the question to the potential carbon and energy sources for these ecosystems. In sedimentary systems, the buried organic matter forms the most obvious feedstock source. Deposited organic matter is exposed to various alteration processes during its geological history and during this maturation pathway, it may release potential substrates for the deep biosphere due to geochemical processes in the diagentic and geothermal processes in the catagenetic phase. Thus, this thesis addresses the availability of potential substrates for deep microbial life being provided by not only geochemically but also geothermally induced alteration processes of terrestrial macromolecular organic matter in lignites and coals of different maturity level. In this context, it was also investigated whether organic carbon rich layers and its adjacent lithologies indeed can form an appropriate habitat for deep microbial communities.

Low molecular weight organic acids (LMWOAs) are suggested to represent important substrates for microbial metabolism. Thus, hydrolysis within the porewater is suggested to be a likely abiotic process releasing formerly kerogen-bound LMWOAs during diagenetic alteration and transformation of the macromolecular organic matter. Therefore, in the current thesis, also the kinetic of the hydrolytic cleavage of ester-bound LMWOAs was investigated in order to gain valuable information on the hydrolytic release of these substrates from lignites and coals at different maturity levels by calculating

11. Conclusions

the reaction rate constants and the activation energies for the hydrolysis reaction and evaluating the effects of the diagenetic and catagenetic maturation of the organic material on these parameters.

Concomitantly, the compositional information obtained during these investigations also allow an important insight into the structural alteration occurring during diagenetic to catagenetic maturation processes within the structure of the macromolecular network forming lignites and coals.

Sample material used for this study

In the scope of this thesis, lignite and coal samples from coal mines and natural outcrops on the North and South Island of New Zealand, being part of the New Zealand coal band, were chosen. The sample material was taken from five different sedimentary basins the Northland, Eastern Southland, Waikato, West Coast and Taranaki Basins, covering diagenetic to catagenetic coalification levels from 0.27% to 0.80% vitrinite reflectance and a time range from Cretaceous to Pleistocene age. In this maturation interval, the biotic and abiotic organic matter alteration is accompanied by the loss of oxygen containing compounds as indicated by the strong decrease of the O/C atomic ratio in the van Krevelen diagram. Thus, the macromolecular organic matter might release oxygen containing organic molecules into the surroundings, being a potential feedstock for deep microbial life. Additionally, sample material from the DEBITS-1 well drilled near the small village of Huntly located in the Waikato coal area on the North Island of New Zealand was chosen. The samples comprise lignites from the Tauranga Group including a transect from an organic carbon rich (lignite) layer into the adjacent organic carbon poor lithologies (silt and sandstones) and sub-bituminous coals from the underlying Te Kuiti Group. The sediments of the Tauranga and the Te Kuiti Groups are separated by an unconformity at about 76 m depth. The Tauranga Group lignites show the same low thermal maturity of 0.29% vitrinite reflectance while the Te Kuiti Group sediments were previously burried deeper to approximately 2000 m and comprise sub-bituminous coals of 0.39% vitrinite reflectance.

The release of low molecular weight organic acids from lignites and coals during diagenetic and catagenetic maturation processes and their implication for the deep biosphere

To investigate the release of kerogen-linked low molecular weight organic acids (LMWOAs) forming a potential feedstock for deep microbial life from lignites and coals during maturation, a novel analytical procedure was developed that enables the detection of LMWOAs within the products of saponification by ion chromatography. Over the chosen maturity range, formate, acetate and oxalate were detected in nearly all samples investigated, showing decreasing amounts with increasing thermal maturity whereas the main decrease is associated with the diagenetic phase reflecting structural alteration processes characterised by the loss of oxygen containing compounds. This phase coincides with temperature conditions being compatible to microbial life in the deep subsurface. The release of these LMWOAs was roughly estimated to be large enough to sustain deep terrestrial microbial life over geological time spans.

Implications from the distribution of high molecular weight organic acids and alcohols linked to the organic macromolecules

Early geochemicaly processes in sedimentary organic matter are characterised by the formation of a complex macromolecular matrix by condensation reactions. The investigation of specific matrix-bound compounds provides, therefore, information about the contributing source material. During ongoing alteration, the macromolecular structure undergoes further alteration processes, including structural reorganisation accompanied by the loss of several compounds.

11. Conclusions

The sources of kerogen-bound alcohols and long chain fatty acids (n-C_{20} to n-C_{30}) are terrestrial plant materials such as coatings and waxes. In contrast, the short chain fatty acids, mainly n-C_{16} and n-C_{18}, may have different sources since they are common in eukaryotes as well as in bacteria where they are part of important cell membrane components. To receive an insight into the contribution and alteration of different source materials including the evidence for the contribution of bacterial biomass the concentrations of kerogen-bound high molecular weight organic acids (HMWOAs) and alcohols were investigated.

The concentrations of ester-bound alcohols show a rapid decrease during diagenesis and remain low during early and main catagenesis. The ester-bound HMWOAs initially also decrease but show an intermittent increase during early catagenesis which is related only to the short chain fatty acids with 16 and 18 carbon atoms before decreasing again during the main catagenesis. The long chain fatty acids with 20 to 30 carbon atoms deriving from terrestrial plant material gradually decrease during diagenesis and early catagenesis and remain constantly low during main catagenesis. It was shown that, especially during the diagenetic phase, the amount of lipids is partly influenced by the contributing organic matter facies. To overcome this problem, the carbon preference index for fatty acids (CPI_{FA}) was introduced. Immature organic matter shows a strong even over odd fatty acid carbon number predominance, which is decreasing with ongoing maturation. Thus, the CPI_{FA} is suggested to represent a less facies-dependent but maturity related signal. The CPI_{FA} for the long chain fatty acids slowly decrease during diagenesis and catagenesis and is close to unity during main catagenesis, reflecting an ongoing maturation trend of the organic matter. For the short chain fatty acids, the CPI_{FA} also initially decreases during diagenesis but reveal an increase during early catagenesis where already increasing amounts of short chain fatty acids were detected. The occurrence of these higher amounts of short chain fatty acids (mainly n-C_{18} and n-C_{18}) with a comparable immature CPI_{FA} signature at this advanced level of maturity might point to the incorporation of more recent biomass. In such deep environments, a conceivable source of immature biomass are deep microbial communities containing C_{16} and C_{18} fatty acids as main cell membrane components. This deep microbial life might have been stimulated by the increasing thermally-induced release of potential substrates from the organic matrix during early catagenesis (*bio-geo coupling hypothesis*).

Comparison of kerogen-bound and bitumen fatty acids

In order to compare the previously investigated kerogen ester linked fatty acids and alcohols with the free fatty acids and alcohols in the bitumen fraction, the lignite and coal maturity series were extracted for their bitumen contents. Differences in the lipid concentrations in the kerogen and bitumen fractions may provide evidence for differences of the contributing source material.

The concentrations of the short and long chain bitumen fatty acids decrease during diagenesis and early catagenesis and were only found in very low amounts in the samples being in the main catagenetic maturation stage. This trend is also reflected by the calculated CPI_{FA} values. The increase of short chain fatty acids during early catagenesis as seen for kerogen bound fraction was not observed. The distribution patterns of the bitumen and the kerogen-bound fatty acids are on a qualitative level similar, but the relative proportion of the short chain to long chain fatty acids is significantly higher in the kerogen-bound fraction, indicating that bitumen and kerogen fraction might not be in a kind of an equilibrium. It is suggested that a higher proportion of microbial biomass (being the remains of microbial communities involved in the transformation of organic carbon rich formations during maturation) might have been directly incorporated into the kerogen matrix after cell death, leading to the higher proportion of the short chain fatty acids in the kerogen.

The kerogen-bound fraction is suggested to be less susceptible to micro-

11. Conclusions

bial degradation during diagenesis and early catagenesis than the bitumen fraction. The ratio of kerogen-bound versus bitumen fatty acids calculated from the lignite and coal samples showed increasing values, revealing this expected lower susceptibility of kerogen-bound fatty acids to degradation in this maturity interval. During ongoing catagenesis, the ratio remains constant or might slightly decrease again, a reason for this being most likely the beginning of thermal degradation of the kerogen.

Low and high molecular weight organic acids in the organic matter rich layers of the DEBITS-1 well and implications for the potential habitat of microbial life from investigation of an organic carbon rich to an organic carbon poor transect.

To receive a deeper insight into the release of substrates (LMWOAs) from organic carbon rich sediments related to maturation and burial depth, samples from lignite and coaly layers from the DEBITS-1 well were investigated.

The DEBITS-1 well penetrates a complex succession of interbedded organic carbon-rich layers and coarser grained mudstones, siltstones and sandstones. The sediments below an unconformity at about 76 m depth forming the Te Kuiti Group were previously buried up to about 2000 m depth and comprise, therefore, sub-bituminous coals with higher thermal maturity (R_0 0.39%) than the sediments of the Tauranga Group located above the unconformity containing lignite layers (R_0 0.29%). Although the samples of the Tauranga Group are mainly of the same thermal maturity, as indicated by vitrinite reflectance data, the relation of the concentrations of kerogen-bound organic acids to burial depth is suggested to provide important information about the alteration process of the organic material, and with this, about the releasing process on a more sensitive and detailed scale. Furthermore, a transect sample set from organic carbon rich (lignite) to organic carbon poor (silt- and sandstones) lithologies was investigated for the concentrations of releasable LMWOAs. The results were compared with permeability and pore space data as well as with microbial lipid data (phospholipids) being indicators for a viable microbial ecosystem and with indiators for fossil microbial ecosystems (ether core lipids) to address the hypothesis that microbial life might inhabit coarser grained organic poor lithologies adjacent to organic carbon rich lithologies (*feeder-carrier hypothesis*).

In the early diagenetic stage lignites from the Tauranga Group, high amounts of kerogen-bound formate, acetate, oxalate as well as propionate and butyrate were detected, whereas the sub-bituminous coals from the underlying Te Kuiti Group contained significantly lower, but still notable, amounts of these LMWOAs. This seems to reflect the previously described maturation related release of these acids, providing a feedstock potential for deep microbial life.

High amounts of short and long chain fatty acids were also detected in the Tauranga Group samples. Although all Tauranga Group samples have the same thermal maturity, at least for the terrigenious long chain fatty acids, the calculated CPI_{FA} values show a strong decreasing trend with burial depth, pointing to ongoing maturation processes with depth in the DEBITS-well. Therefore, the trend outlined by the CPI_{FA} for the long chain fatty acids seems to reflect the maturation of the organic matter on a more detailed and smaller scale. The short chain fatty acids reveal some scattering also in the CPI_{FA} data, suggesting a stronger influence by non maturity related factors (for example, the contribution of short chain fatty acids from bacterial biomass). The concentrations and CPI_{FA} values of the Te Kuiti Group coals are lower and less variable, which is in accordance with their higher level of maturity.

Significant amounts of kerogen-bound LMWOAs were also found in the transect samples in decreasing concentrations with increasing distance from the lignite layer. The phospholipids (PLs) being indicators for viable microbial life show highest abundances in the more permeable organic carbon poor silt- and sandstones adjacent to the organic carbon rich layer, suggesting that microbial communities are located in these layers and being fed by

11. Conclusions

substrates released from the lignite layer into the adjacent lithologies. As discussed above, there is also evidence for microbial life within the organic matrix of the lignite layers providing a huge excess of potential substrates. A reason why PLs are not detected in the organic carbon rich layer is provided by the observation that the detection of PLs might be simply suppressed in samples with a high background signal of organic matter in the electrospray (ESI) interface during measurement. However, the activity and abundance of microbial life might be higher in the more permeable lithologies.

Hydrolytic ester cleavage as potential process to release substrates from organic matter rich lithologies - implications from kinetic investigations

The most favorable process releasing the LMWOAs into the surrounding is hydrolysis within the porewater. This process might regulate the abiotic release of the substrates from the organic carbon rich lithologies. For a deeper understanding of this process, the kinetic parameters such as the reaction rate constants (k'), the activation energies (E_A) and the corresponding frequency factors (A) were determined from selected lignite and coal samples of different maturity (two samples each from different diagenetic and catagenetic maturation levels). The determined values also provide a deeper insight into the stability of the ester-bound LMWOAs in relation to the maturity of the macromolecular organic material and its structural properties influenced by the reorganisation process of the organic macromolecular matrix during alteration.

Although the exact values determined for the reaction rate constants k', the activation energies E_A and the frequency factors A are to some extend questionable due to experimental problems, the determined data provide interesting insights into general trends of ester hydrolysis in organic matter of different maturity. The experimental determined k' for formate and acetate decreases with ongoing maturation, pointing to a slower cleavage reaction in the more mature samples. Concomitantly, the E_A values and the frequency factors A increase with ongoing maturation, reflecting a higher energy demand and a higher number of collisions needed for the reaction for the cleavage of esters in the mature samples. This suggests that ester bonds within the more mature kerogen are better sterically protected by the compact macromolecular structure. Despite the fact that the hydrolysis experiements were carried out at relatively high temperatures compared to the environmental conditions within the sediments (at least during the early diagenetic phase) the half life's ($t_{1/2}$) calculated for hydrolysis by the porewater at approximately pH 7 (ranging from approximately 100 to several 10000 years) appear to be significantly too low to explain the observed release of LMWOAs from the coal samples during millions of years. This suggests that additional factors (e.g. pore space, permeability, pore water flow and diffusion) might influence and slow down the LMWOA release within natural environments. Nevertheless, the observed increase of the calculated half lifes with ongoing maturation also reflects the slower LMWOA release in the more mature samples.

Insights into the network structure of the organic macromolecule from BBr_3 ether cleavage products.

The organic macromolecular matrix of lignites and coals is a complex network in which aromatic structural units deriving e.g. from lignin of woody tissue are linked either directly or by aliphatic chains. In addition to ester functions, another important linkage, therein, is provided by ether functions. Therefore, important cross-linking units are aliphatic alcohols with more than one hydroxy groups. Alcohols with only one hydroxy group are suggested to derive from terrestrial plant waxes and coatings and are terminally linked to the organic matrix. To receive a deeper insight into the complex structure of the macromolecules and the maturation related changes within the network structure, ether cleavage was applied to the lignite and coal samples of different maturity.

11. Conclusions

Terminal ether-bound alcohols were detected in strongly decreasing concentrations related to thermal maturation indicating a major loss of these compounds during diagenesis. Although ether bonds are known to be more stable than ester bonds, their main decomposition is observed in the same maturation range and it is speculated that the stability differences between both linkages are not strong enough to be observed in alteration processes over geological time scales. Ether-bound alcohols with more than one hydroxy group were found with chain length up to five carbon atoms representing important cross-linking structures within the kerogen matrix. Their different occurrence suggest a high dependence to the contributing plant facies. These compounds were found in relatively high concentrations, even in the main catagenetic stage samples, suggesting that these linkage structures are sterically protected within the macromolecular network during the maturation process. Together with the short chain ester-bound moieties (such as oxalic acid), these cross-linking units appear to be of higher relevance than the cross-linking long chain dicarboxylic acids which were only found in the early diagenetic stage lignites after saponification. Additional cross-linking substructures were poly-hydroxy aromatics, hydroxy esters and ketones.

Addittionally, fatty acid ethyl esters were detected, representing formerly kerogen-bound fatty acids deriving most likely from higher terrestrial plant input. Their amounts are in the same order of magnitude as the bitumen fatty acids and the kerongen-bound fatty acids released by saponification. This, and the fact that they show a different distribution pattern than the kerogen-bound fatty acid fraction obtained after saponification suggests that they represent an inner kerogen fatty acid fraction which is not accessible by the alkaline ester cleavage procedure.

References

Adler, E. (1977). Lignin chemistry - past, present and future. *Wood Science Technology*, 11:69–218.

Albrecht, P., Vandenbroucke, M., and Mandengué, M. (1976). Geochemical studies on the organic matter from the Douala Basin (Cameroon) - I. Evolution of the extractable organic matter and the formation of petroleum. *Geochimica et Cosmochimica Acta*, 40(7):791–799.

Allan, J., Bjorøy, M., and Douglas, A. (1977). Variation in the content and distribution of high molecular weight hydrocarbons in a series of coal macerals of different ranks. *Advances in Organic Geochemistry 1975*, pages 633–654.

Allan, P. A. and Allen, J. R. (1998). *Basin Analysis - Principles and Applications*. Blackwell Science KK, Oxford.

Allibone, A. H. and Tulloch, A. J. (1997). Metasedimentary, granitoid, and gabbroic rocks from central Stewart Island, New Zealand. *New Zealand Journal of Geology and Geophysics*, 40(1):422–432.

Ambles, A., Grasset, L., Dupas, G., and Jacquesy, J. C. (1996). Ester- and ether bond cleavage in immature kerogens. In *17th International Meeting on Organic Geochemistry*, pages 681–690, Donostia-San Sebastian, Spain.

Amblès, A., Halim, M., Jacquesy, J., Vitorovic, D., and Ziyad, M. (1994). Characterization of kerogen from Timahdit shale (Y-layer) based on multistage alkaline permanganate degradation. *Fuel*, 73:17–24.

References

Arnosti, C. and Jørgensen, B. B. (2003). High activity and low temperature optima of extracellular enzymes in Arctic sediments: Implications for carbon cycling by heterotrophic microbial communities. *Marine Ecology Progress Series*, 249:15–24.

Arnosti, C., Jørgensen, B. B., Sagemann, J., and Thamdrup, B. (1998). Temperature dependance of microbial degradation of organic matter in marine sediments: polysaccharide hydrolysis, oxygen consumption, and sulfate reduction. *Marine Ecology Progress Series*, 165:59–70.

Asperger, A., Engewald, W., and Fabian, G. (1999). Analytical characterization of natural waxes employing pyrolysis-gas chromatography-mass spectrometry. *Journal of Analytical and Applied Pyrolysis*, 50(2):103–115.

Atkins, P. W. (1986). *Physical Chemistry*. Oxford University Press, Oxford, 3rd edition.

Baas, M., Pancost, R., van Geel, B., and Sinninghe Damsté, J. S. (2000). A comparative study of lipids in sphagnum species. *Organic Geochemistry*, 31(6):535–541.

Baas-Becking, L. G. M., van Stockum, W. P., and Zoon, N. V. (1934). *Geobiologie of Inleiding Tot de Milieukunde*. The Hague, The Netherlands.

Back, T. G., Yang, K., and Krouse, H. R. (1992). Desulfurization of benzo- and dibenzothiophenes with nickel boride. *Journal of Organic Chemistry*, 57(7):1986–1990.

Bada, J. L. and Mann, E. H. (1980). Amino acid diagenesis in DSDP cores: Kinetics and mechanisms of some reactions and their applications in geochronology and in paleotemperature and heat flow determinations. *Earth Science Reviews*, 16:21–57.

Barakat, A. O. and Rullkötter, J. (1995). Extractable and bound fatty acids in core sediments from the Nördlinger Ries, southern Germany. *Fuel*, 74(3):416–425.

Barry, J. M., Duff, S. W., and Macfarlan, D. A. B. (1994). Coal resources of New Zealand. *Resource information report*, 16.

Barth, T., Borgund, A. E., Hopland, A. L., and Graue, A. (1988). Volatile organic acids produced during kerogen maturation - amounts, composition and role in migration of oil. *Organic Geochemistry*, 13(1-3):461–465.

Bates, A. L. and Hatcher, P. G. (1989). Solid-state ^{13}C NMR studies of a large fossil gymnosperm from the Yallourn Open Cut, Latrobe Valley, Australia. *Organic Geochemistry*, 14(6):609–617.

Bechtel, A., Sachsenhofer, R., Zdravkov, A., Kostova, I., and Gratzer, R. (2005). Influence of floral assemblage, facies and diagenesis on petrography and organic geochemistry of the Eocene Bourgas coal and the Miocene Maritza-East lignite (Bulgaria). *Organic Geochemistry*, 36:1498–1522.

Bend, S. L. (1992). The origin, formation and petrographic composition of coal. *Fuel*, 71:851–870.

Benkeser, R. A., Robinson, R. E., Sauve, D. M., and Thomas, O. H. (1955). Reduction of organic compounds by lithium in low molecular weight amines. I. Selective reduction of aromatic hydrocarbons to monoolefins. *Journal of the American Chemical Society*, 77(12):3230–3233.

Bennett, B., Bowler, B. F. J., and Larter, S. (1996). Determination of C_0-C_3 alkylphenols in crude oils and waters. *Analytical Chemistry*, 68:3697–3702.

Berner, R. A. (1984). Sedimentary pyrite formation: An update. *Geochimica et Cosmochimica Acta*, 48(4):605–615.

Berner, R. A. (1994). GEOCARB II: A revised model of atmospheric CO_2 over Phanerozoic time. *American Journal of Science*, 294(1):56–91.

Béhar, F. (1997). Thermal cracking of kerogen in open and closed systems: Determination of kinetic parameters and stoichiometric coefficients for oil and gas generation. *Organic Geochemistry*, 26(5-6):321–339.

References

Béhar, F., Beaumont, V., and De Barros-Penteado, H. L. (2001). Rock-Eval 6 technology: Performances and developements. Oil & Gas Science and Technology. *Revue de l'Institut Francais du Pétrole*, 56:111–134.

Béhar, F., Gillaizeau, B., Derenne, S., and Largeau, C. (2000). Nitrogen distribution in the pyrolysis products of a type II kerogen (Cenomanian, Italy). Timing of molecular nitrogen production versus other gases. *Energy & Fuels*, 14(2):431–440.

Béhar, F. and Vandenbroucke, M. (1987). Chemical modelling of kerogens. *Organic Geochemistry*, 11(1):15–24.

Béhar, F., Vandenbroucke, M., Teermann, S. C., Hatcher, P. G., Leblond, C., and Lerat, O. (1995). Experimental simulation of gas generation from coals and a marine kerogen. *Chemical Geology*, 126(3-4):247–260.

Bhatt, M. V. and Kulkarni, S. U. (1983). Cleavage of ethers. *Synthesis*, 4:249–282.

Blake, E. S., Hammann, W. C., Edwards, J. W., Reichard, T. E., and Ort, M. R. (1961). Thermal stability as a function of chemical structure. *Journal of Chemical & Engineering Data*, 6(1):87–98.

Blau, K. and Darbre, A. (1993). Thionyl chloride in ester formation. In Blau, K. and Halket, J., editors, *Handbook of Derivates for Chromatography*, page 16. John Wiley and Sons, Chichester, 2nd edition.

Blöchl, E., Rachel, R., Burggraf, S., Hafenbradl, D., Jannasch, H. W., and Stetter, K. O. (1997). Pyrolobus fumarii, gen. and sp. nov., represents a novel group of archea, extending the upper temperature limit for life to 113 degree C. *Extremophiles*, 1:14–21.

Blodau, C. (2006). A review of acidity generation and consumption in acidic coal mine lakes and their watersheds. *Science of The Total Environment*, 369(1-3):307–332.

Blokker, P., Schouten, S., de Leeuw, J. W., Sinninghe Damsté, J. S., and van den Ende, H. (2000). A comparative study of fossil and extant algaenans using ruthenium tetroxide degradation. *Geochimica et Cosmochimica Acta*, 64(12):2055–2065.

Blokker, P., Schouten, S., van den Ende, H., de Leeuw, J. W., Hatcher, P. G., and Sinninghe Damsté, J. S. (1998). Chemical structure of algaenans from the fresh water algae *Tetraedron minimum*, *Scenedesmus communis* and *Pediastrum boryanum*. *Organic Geochemistry*, 29(5-7):1453–1468.

Blokker, P., van Bergen, P., Pancost, R., Collinson, M. E., de Leeuw, J. W., and Sinninghe Damsté, J. S. (2001). The chemical structure of Gloeocapsomorpha prisca microfossils: Implications for their origin. *Geochimica et Cosmochimica Acta*, 65(6):885–900.

Bou-Raad, M., Hobday, M. D., and Rix, C. J. (2000). Aqueous extraction of oxalate and other anions from coal. *Fuel*, 79(10):1185–1193.

Bradshaw, J. D. (1989). Cretaceous geotectonic patterns in the New Zealand region. *Tectonics*, 8(4):803–820.

Bradshaw, J. D., Pankhurst, R. J., Weaver, S. D., Storey, B. C., Muir, R. J., and Ireland, T. R. (1997). New Zealand superterranes recognized in Mary Burd Land and Thurston Island. *Terra Antarctica*, 3:429–436.

Bray, E. E. and Evans, E. D. (1961). Distribution of n-paraffins as a clue to recognition of source beds. *Geochimica et Cosmochimica Acta*, 22(1):2–15.

Brown, J., Kasrai, M., Bancroft, G., Tan, K., and Ghen, J. (1992). Direct identification of organic sulphur species in Rasa coal from sulphur L-edge X-ray absorption near-edge spectra. *Fuel*, 71(6):649–653.

Burgess, C. E. and Schobert, H. H. (1998). Relationship of coal characteristics determined by pyrolysis/gas chromatography/mass spectrometry and nuclear magnetic resonance to liquefaction reactivity and product composition. *Energy & Fuels*, 12(6):1212–1222.

References

Burnham, A. K., Braun, R. L., Gregg, H. R., and Samoun, A. M. (1987). Comparison of methods for measuring kerogen pyrolysis rates and fitting kinetic parameters. *Energy & Fuels*, 1:452–458.

Caldin, E. F. and Long, G. (1954). The equilibrium between ethoxide and hydroxide ions in ethanol and in ethanol-water mixtures. *Journal of Chemical Society*, pages 3737–3742.

Carlsen, P. H. J., Katsuki, T., Martin, V. S., and Sharpless, K. B. (1981). A greatly improved procedure for ruthenium tetroxide catalyzed oxidations of organic compounds. *Journal of Organic Chemistry*, 46(19):3936–3938.

Castenholz, R. W. (1996). Endemism and biodiversity of thermophilic cyanobacteria. *Nova Hedwigia Beiheft*, 112:33–47.

Chappe, B., Michaelis, W., and Albrecht, P. (1982). Polar lipids of archaebacteria in sediments and petroleums. *Science*, 217:65–66.

Charrié-Duhaut, A., Lemoine, S., Adam, P., Connan, J., and Albrecht, P. (2000). Abiotic oxidation of petroleum bitumens under natural conditions. *Organic Geochemistry*, 31(10):977–1003.

Chen, Y., Senesi, N., and Schnitzer, M. (1977). Information provided on humic substances by E4/E6 ratios. *Soil Science Society of America Journal*, 41(2):352–358.

Chivian, D., Brodie, E. L., Alm, E. J., Culley, D. E., Dehal, P. S., DeSantis, T. Z., Gihring, T. M., Lapidus, A., Lin, L.-H., Lowry, S. R., Moser, D. P., Richardson, P. M., Southam, G., Wanger, G., Pratt, L. M., Andersen, G. L., Hazen, T. C., Brockman, F. J., Arkin, A. P., and Onstott, T. C. (2008). Environmental genomics reveals a single-species ecosystem deep within Earth. *Science*, 322(5899):275–278.

Cho, J.-C. and Tiedje, J. M. (2000). Biogeography and degree of endemicity of fluorescent Pseudomonas strains in soil. *Applied and Environmental Microbiology*, 66(12):5448–5456.

Christensen, B., Laake, M., and Lien, T. (1996). Treatment of acid mine water by sulfate-reducing bacteria; results from a bench scale experiment. *Water Research*, 30(7):1617–1624.

Christl, I., Knicker, H., Kögel-Knabner, I., and Kretzschmar, R. (2000). Chemical heterogeneity of humic substances: Characterization of size fractions obtained by hollow-fibre ultrafiltration. *European Journal of Soil Science*, 51(4):617–625.

Collins, M. J., Westbroek, P., Muyzer, G., and de Leeuw, J. W. (1992). Experimental evidence for condensation reactions between sugars and proteins in carbonate skeletons. *Geochimica et Cosmochimica Acta*, 56(4):1539–1544.

Combaz, A. and de Matharel, M. (1978). Organic sedimentation and genesis of petroleum in Mahakam Delta, Borneo. *AAPG Bulletin*, 62(9):1684–1695.

Connan, J. (1984). Biodegradation in crude oils in reservoirs. In Brooks, J. and Welte, D., editors, *Advances in Petroleum Geochemistry*, volume 1, pages 299–355. Academic Press, London.

Cooles, G. P., Mackenzie, A. S., and Parkes, R. J. (1987). Non-hydrocarbons of significance in petroleum exoploration: Volatile fatty acids and non-hydrocarbon gases. *Mineralogical Magazine*, 51:483–493.

Cooper, J. E. (1962). Fatty acids in recent and ancient sediments and petroleum reservoir waters. *Nature*, 193(4817):744–746.

Cooper, R. A. and Tulloch, A. J. (1992). Early Palaeozoic terranes in New Zealand and their relationship to the Lachlan Fold Belt. *Tectonophysics*, 214(1-4):129–144.

Cowan, D. A. (2004). The upper temperature for life - where do we draw the line? *Trends in Microbiology*, 12(2):58–60.

References

Cranwell, P. A. (1974). Monocarboxylic acids in lake sediments: Indicators, derived from terrestrial and aquatic biota, of paleoenvironmental trophic levels. *Chemical Geology*, 14(1-2):1–14.

Cranwell, P. A. (1977). Organic geochemistry of Can Loch (Sutherland) sediments. *Chemical Geology*, 20:205–211.

Day, J. N. E. and Ingold, C. K. (1941). Mechanism and kinetics of carboxylic ester hydrolysis and carboxyl esterification. *Transactions of the Faraday Society*, 37:686–705.

DeLong, E. F. (2004). Microbiology: Microbial life breathes deep. *Science*, 306(5705):2198–2200.

Demaison, G. J. and Moore, G. T. (1980). Anoxic environments and oil source bed genesis. *AAPG Bulletin*, 64(8):1179–1209.

Derbyshire, F., Marzec, A., Schulten, H.-R., Wilson, M. A., Davis, A., Tekley, P., Delpuech, J.-J., Jurkiewicz, A., Bronnimann, C. E., Wind, R. A., Maciel, G. E., Narayan, R., Bartle, K., and Snape, C. (1989). Molecular structure of coals: A debate. *Fuel*, 68:1091–1106.

Devine, W. T. (1982). Nature conservation and land-use history of the Chatham Islands, New Zealand. *Biological Conservation*, 23(2):127–140.

D'Hondt, S., Jørgensen, B. B., Miller, D. J., Batzke, A., Blake, R., Cragg, B. A., Cypionka, H., Dickens, G. R., Ferdelman, T., Hinrichs, K.-U., Holm, N. G., Mitterer, R., Spivack, A., Wang, G., Bekins, B., Engelen, B., Ford, K., Gettemy, G., Rutherford, S. D., Sass, H., Skilbeck, C. G., Aiello, I. W., Guerin, G., House, C. H., Inagaki, F., Meister, P., Naehr, T., Niitsuma, S., Parkes, R. J., Schippers, A., Smith, D. C., Teske, A., Wiegel, J., Padilla, C. N., and Acosta, J. L. S. (2004). Distributions of microbial activities in deep subseafloor sediments. *Science*, 306(5705):2216–2221.

D'Hondt, S., Rutherford, S., and Spivack, A. J. (2002). Metabolic activity of subsurface life in deep-sea sediments. *Science*, 295(5562):2067–2070.

di Primio, R. and Horsfield, B. (1996). Predicting the generation of heavy oils in carbonate/evaporitic environments using pyrolysis methods. *Organic Geochemistry*, 24:999–1016.

Di Primio, R., Horsfield, B., and Guzman-Vega, M. A. (2000). Determining the temperature of petroleum formation from the kinetic properties of petroleum asphaltenes. *Nature*, 406:173–176.

Dieckmann, V., Schenk, H. J., Horsfield, B., and Welte, D. H. (1998). Kinetics of petroleum generation and cracking by programmed-temperature closed-system pyrolysis of toarcian shales. *Fuel*, 77(1-2):23–31.

Disnar, J.-R., Stefanova, M., Bourdon, S., and Laggoun-Défarge, F. (2005). Sequential fatty acid analysis of a peat core covering the last two millennia (Tritrivakely lake, Madagascar): Diagenesis appraisal and consequences for palaeoenvironmental reconstruction. *Organic Geochemistry*, 36(10):1391–1404.

Djerassi, C. and Engle, R. R. (1953). Oxidations with ruthenium tetroxide. *Journal of the American Chemical Society*, 75(15):3838–3840.

Down, A. L. and Himus, G. W. (1941). A preliminary study of the chemical constitution of kerogen. *Journal of the Institute of Petroleum*, 27:426–445.

Durand, B. (1980). Sedimentary organic matter and kerogen. definition and quantitative importance of kerogen. In Durand, B., editor, *Kerogen*, pages 13–14. Technip, Paris.

Durand, B. and Espitalié, J. (1973). Evolution de la matière organique au cours de l'enfouissement des sediments. *Compte Rendus de l'Académie des Sciences (Paris)*, 276:2253–2256.

Durand, B. and Espitalie, J. (1976). Geochemical studies on the organic matter from the Douala Basin (Cameroon) - II. Evolution of kerogen. *Geochimica et Cosmochimica Acta*, 40(7):801–808.

References

Durand, B., Nicaise, G., Roucaché, J., Vandenbroucke, M., and Hagemann, H. (1977). Etude géochimique d'une série de charbons. In Campos, R. and Goni, J., editors, *Advances in Organic Geochemistry 1975*, pages 601–631. ENADIMSA, Madrid.

Edbrooke, S. W., Sykes, R., and Pocknall, D. T. (1994). Geology of the Waikato Coal Measures, Waikato coal region, New Zealand. *Institute of Geological & Nuclear Sciences monograph. Lower Hutt, New Zealand: Institute of Geological and Nuclear Sciences Limited.* 6, page 236 p.

Eglinton, G. and Hamilton, R. J. (1967). Leaf epicuticular waxes. *Science*, 156(3780):1322–1335.

Eglinton, G., Logan, G. A., Ambler, R. P., Boon, J. J., and Perizonius, W. R. K. (1991). Molecules through time: Fossil molecules and biochemical systematics. *Philosophical Transactions: Biological Sciences*, 333(1268):315–328.

Eglinton, T. I., Irvine, J. E., Vairavamurthy, A., Zhou, W., and Manowitz, B. (1994). Formation and diagenesis of macromolecular organic sulfur in Peru margin sediments. *Organic Geochemistry*, 22(3-5):781–799.

Eglinton, T. I., Sinninghe Damsté, J. S., Pool, W., de Leeuw, J. W., Eijk, G., and Boon, J. J. (1992). Organic sulphur in macromolecular sedimentary organic matter. II. Analysis of distributions of sulphur-containing pyrolysis products using multivariate techniques. *Geochimica et Cosmochimica Acta*, 56(4):1545–1560.

Ericsson, I. and Lattimer, R. P. (1989). Pyrolysis nomenclature. *Journal of Analytical and Applied Pyrolysis*, 14(4):219–221.

Espitalié, J., Deroo, G., and Marquis, F. (1985). Rock-Eval pyrolysis, and its applications. *Revue de l'Institut Français du Pétrole*, 40:563–579.

Exner, O. (1988). Why are carboxylic acids and phenols stronger acids than alcohols? *Journal of Organic Chemistry*, 53(8):1810–1812.

Falkowski, P. G. (1997). Evolution of the nitrogen cycle and its influence in the biological sequestration of CO_2 in the ocean. *Nature*, 387:272–275.

Falkowski, P. G., Fenchel, T., and Delong, E. F. (2008). The microbial engines that drive Earth's biogeochemical cycles. *Science*, 320:1034–1039.

Fang, J., Barcelona, M. J., Nogi, Y., and Kato, C. (2000). Biochemical implications and geochemical significance of novel phospholipids of the extremely barophilic bacteria from the Marianas Trench at 11,000 m. *Deep Sea Research*, 47:1173–1182.

Farrington, J. W. and Quinn, J. G. (1971). Comparison of sampling and extraction techniques for fatty acids in recent sediments. *Geochimica et Cosmochimica Acta*, 35(7):735–741.

Farrington, J. W. and Quinn, J. G. (1973). Biogeochemistry of fatty acids in recent sediments from Narragansett Bay, Rhode Island. *Geochimica et Cosmochimica Acta*, 37(2):259–268.

Fester, J. I. and Robinson, W. E. (1966). Oxygen functional groups in Green River Shale oil-shale kerogen and trona acids. In Gould, R. F., editor, *Coal Science*, pages 22–31. American Chemical Society, Washington, DC.

Forsman, J. and Hunt, J. (1958). Insoluble organic matter (kerogen) in sedimentary rocks. *Geochimica et Cosmochimica Acta*, 15(3):170–182.

Franke, R. and Schreiber, L. (2007). Suberin - a biopolyester forming apoplastic plant interfaces. *Current Opinion in Plant Biology*, 10(3):252–259.

Fredrickson, J. K., McKinley, J. P., Bjornstad, B. N., Long, P. E., Ringelberg, D. B., White, D. C., Krumholz, L. R., Suflita, J. M., Colwell, F. S., Lehman, R. M., and Phelps, T. J. (1997). Pore-size constraints on the activity and survival of subsurface bacteria in a late Cretaceous shale-sandstone sequence, northwestern New Mexico. *Geomicrobiology Journal*, 14:183–202.

References

Fredrickson, J. K. and Onstott, T. C. (1996). Microbes deep inside the Earth. *Scientific American*, October:42–47.

Fredrickson, J. K. and Onstott, T. C. (2001). Biogeochemical and geological significance of subsurface microbiology. In Fredrickson, J. K. and Fletcher, M., editors, *Subsurface Microbiology and Biogeochemistry*, pages 3–37. Wiley-Liss, Inc.

Froelich, P. N., Klinkhammer, G. P., Bender, M. L., Luedtke, N. A., Heath, G. R., Cullen, D., Dauphin, P., Hammond, D., Hartman, B., and Maynard, V. (1979). Early oxidation of organic matter in pelagic sediments of the eastern equatorial Atlantic: Suboxic diagenesis. *Geochimica et Cosmochimica Acta*, 43(7):1075–1090.

Gilichinsky, D. and Wagener, S. (1995). Microbial life in permafrost: A historical review. *Permafrost and Periglacial Processes*, 6(3):243–250.

Gillaizeau, B., Béhar, F., Derenne, S., and Largeau, C. (1997). Nitrogen fate during laboratory maturation of a type I kerogen (Oligocene, Turkey) and related algaenan: Nitrogen mass balances and timing of N_2 production versus other gases. *Energy & Fuels*, 11(6):1237–1249.

Given, P. H., Spackman, W., Painter, P. C., Rhoads, C. A., Ryan, N. J., Alemany, L., and Pugmire, R. J. (1984). The fate of cellulose and lignin in peats: An exploratory study of the input to coalification. *Organic Geochemistry*, 6:399–407.

Glombitza, C., Mangelsdorf, K., and Horsfield, B. (2009a). Maturation related changes in the distribution of ester bound fatty acids and alcohols in a coal series from the New Zealand Coal Band covering diagenetic to catagenetic coalification levels. *Organic Geochemistry*, 40(10):1063–1073.

Glombitza, C., Mangelsdorf, K., and Horsfield, B. (2009b). A novel procedure to detect low molecular weight compounds released by alkaline ester cleavage from low maturity coals to assess its feedstock potential for deep microbial life. *Organic Geochemistry*, 40(2):175–183.

Gülz, P.-G. (1994). Epicuticular leaf waxes in the evolution of the plant kingdom. *Journal of Plant Physiology*, 143:453–464.

Grasset, L., Guignard, C., and Amblès, A. (2002). Free and esterified aliphatic carboxylic acids in humin and humic acids from a peat sample as revealed by pyrolysis with tetramethylammonium hydroxide or tetraethylammonium acetate. *Organic Geochemistry*, 33(3):181–188.

Gütlich, P. (1970). Physikalische Methoden in der Chemie: Mössbauer-Spektroskopie I. *Chemie in unserer Zeit*, 4(5):133–144.

Guckert, J. B., Hood, M. A., and White, D. C. (1986). Phospholipid ester-linked fatty acid profile changes during nutrient deprivation of *Vibrio cholerae*: Increases in the trans/cis ratio and proportions of cyclopropyl fatty acids. *Applied and Environmental Microbiology*, 52:794–801.

Harvey, H. R., Fallon, R. D., and Patton, J. S. (1986). The effect of organic matter and oxygen on the degradation of bacterial membrane lipids in marine sediments. *Geochimica et Cosmochimica Acta*, 50(5):795–804.

Hatcher, P. (1990). Chemical structural models for coalified wood (vitrinite) in low rank coal. *Organic Geochemistry*, 16(4-6):959–968.

Hatcher, P., Breger, I., Szeverenyi, N., and Maciel, G. (1982). Nuclear magnetic resonance studies of ancient buried wood - II. Observations on the origin of coal from lignite to bituminous coal. *Organic Geochemistry*, 4(1):9–18.

Hatcher, P. G. (1988). Dipolar-dephasing ^{13}C NMR studies of decomposed wood and coalified xylem tissue: Evidence for chemical structural changes associated with defunctionalization of lignin structural units during coalification. *Energy & Fuels*, 2(1):48–58.

Hatcher, P. G. and Clifford, D. J. (1997). The organic geochemistry of coal: From plant materials to coal. *Organic Geochemistry*, 27(5-6):251–274.

References

Hatcher, P. G., Faulon, J. L., Wenzel, K. A., and Cody, G. D. (1992). A structural model for lignin-derived vitrinite from high-volatile bituminous coal (coalified wood). *Energy & Fuels*, 6(6):813–820.

Hatcher, P. G., Lerch, H. E., Bates, A. L., and Verheyen, T. V. (1989). Solid-state ^{13}C nuclear magnetic resonance studies of coalified gymnosperm xylem tissue from Australian brown coals. *Organic Geochemistry*, 14(2):145–155.

Head, I. M., Jones, D. M., and Larter, S. R. (2003). Biological activity in the deep subsurface and the origin of heavy oil. *Nature*, 426(6964):344–352.

Hedges, J. I. (1978). The formation and clay mineral reactions of melanoidins. *Geochimica et Cosmochimica Acta*, 42(1):69–76.

Hedges, J. I., Cowie, G. L., Ertel, J. R., James Barbour, R., and Hatcher, P. G. (1985). Degradation of carbohydrates and lignins in buried woods. *Geochimica et Cosmochimica Acta*, 49:701–711.

Höld, I. M., Brussee, N. J., Schouten, S., and Sinninghe Damsté, J. S. (1998). Changes in the molecular structure of a type II-S kerogen (Monterey Formation, U.S.A.) during sequential chemical degradation. *Organic Geochemistry*, 29(5-7):1403–1417.

Hofmann, I. C., Hutchison, J., Robson, J. N., Chicarelli, M. I., and Maxwell, J. R. (1992). Evidence for sulphide links in a crude oil asphaltene and kerogens from reductive cleavage by lithium in ethylamine. *Organic Geochemistry*, 19(4-6):371–387.

Hopmans, E. C., Weijers, J. W. H., Schefuß, E., Herfort, L., Sinninghe Damsté, J. S., and Schouten, S. (2004). A novel proxy for terrestrial organic matter in sediments based on branched and isoprenoid tetraether lipids. *Earth and Planetary Science Letters*, 224(1-2):107–116.

Horneck, G. (2000). The microbial world and the case for Mars. *Planetary and Space Science*, 48(11):1053–1063.

Horsfield, B. (1984). Pyrolysis studies and petroleum exploration. In Brooks, J. and Welte, D. H., editors, *Advances in Petroleum Geochemistry*, volume 1, pages 247–292. Academic Press, New York.

Horsfield, B. (1989). Practical criteria for classifying kerogens: Some observations from pyrolysis-gas chromatography. *Geochimica et Cosmochimica Acta*, 53(4):891–901.

Horsfield, B., Disko, U., and Leistner, F. (1989). The micro-scale simulation of maturation: Outline of a new technique and its potential applications. *Geologische Rundschau*, 78:361–374.

Horsfield, B., Heckers, J., Leythaeuser, D., Littke, R., and Mann, U. (1991). A study of the Holzener Asphaltkalk, northern Germany: Observations regarding the distribution, composition and origin of organic matter in an exhumed petroleum reservoir. *Marine and Petroleum Geology*, 8:198–211.

Horsfield, B., Kieft, T., Amann, H., Franks, S., Kallmeyer, J., Mangelsdorf, K., Parkes, R. J., Wagner, W., Wilkes, H., and Zink, K. G. (2007). The geobiosphere. In Harms, U., Koeberl, C., and Zoback, M. D., editors, *Continental Scientific Drilling: A Decade of Progress and Challenges for the Future*, pages 163–211. Springer, Berlin- Heidelberg.

Horsfield, B., Schenk, H. J., Zink, K., Ondrak, R., Dieckmann, V., Kallmeyer, J., Mangelsdorf, K., di Primio, R., Wilkes, H., Parkes, R. J., Fry, J., and Cragg, B. (2006). Living microbial ecosystems within the active zone of catagenesis: Implications for feeding the deep biosphere. *Earth and Planetary Science Letters*, 246(1-2):55–69.

Huc, A., Durand, B., Roucachet, J., Vandenbroucke, M., and Pittion, J. (1986). Comparison of three series of organic matter of continental origin. *Organic Geochemistry*, 10(1-3):65–72.

Huffman, G. P., Mitra, S., Huggins, F. E., Shah, N., Vaidya, S., and Lu, F. (1991). Quantitative analysis of all major forms of sulfur in coal by X-ray absorption fine structure spectroscopy. *Energy & Fuels*, 5(4):574–581.

References

Huffman, G. P., Shah, N., Huggins, F. E., Stock, L. M., Chatterjee, K., Kilbane, J. J., Chou, M.-I. M., and Buchanan, D. H. (1995). Sulfur speciation of desulfurized coals by XANES spectroscopy. *Fuel*, 74(4):549–555.

Hutton, A. C. (1987). Petrographic classification of oil shales. *International Journal of Coal Geology*, 8(3):203–231.

Hvoslef, S., Larter, S., and Leythaeuser, D. (1988). Aspects of generation and migration of hydrocarbons from coal-bearing strata of the Hitra formation, Haltenbanken area, offshore Norway. *Organic Geochemistry*, 13:525–536.

ICCP (1963). *International Handbook of Coal Petrology*. Centre National de la Recherche Scientifique, Paris, France, 2^{ed} edition.

Isaac, M. J., Herzer, R. H., Brook, F. J., and Hayward, B. W. (1994). Cretaceous and Cenozoic sedimentary basins of Northland, New Zealand. *Institute of Geological & Nuclear Sciences monograph*. Lower Hutt, New Zealand: Institute of Geological and Nuclear Sciences Limited, 8:203 pp.

Isaac, M. J. and Lindqvist, J. K. (1990). Geology of lignite resources of the East Southland Group, New Zealand. *New Zealand Geological Survey Bulletin*, 101:202 pp.

Ives, D. A. J. and O'Neill, A. N. (1958). The chemistry of peat: II. The triterpenes of peat moss (sphagnum). *Canadian Journal of Chemistry*, 36(6):926–930.

Jacobs, P. and Cnudde, V. (2009). Applications of X-ray computed tomography in engineering geology or looking inside rocks. *Engineering Geology*, 103(3-4):67–68.

Jonkers, H. M., Ludwig, R., Wit, R., Pringault, O., Muyzer, G., Niemann, H., Finke, N., and Beer, D. (2003). Structural and functional analysis of a microbial mat ecosystem from a unique permanent hypersaline inland lake: La Salada de Chiprana (NE Spain). *FEMS Microbiology Ecology*, 44(2):175–189.

Jørgensen, B. B. (1982). Mineralization of organic matter in the sea bead - the role of sulfate reduction. *Nature*, 296:643–645.

Jørgensen, B. B., Isaksen, M. F., and Jannasch, H. W. (1992). Bacterial sulfate reduction above 100°C in deep-sea hydrothermal vent sediments. *Science*, 258:1756–1757.

Kallmeyer, J., Mangelsdorf, K., Cragg, B., and Horsfield, B. (2006). Techniques for contamination assessment during drilling for terrestrial subsurface sediments. *Geomicrobiology Journal*, 23(3-4):227–239.

Kashefi, K. (2004). Response to Cowan: The upper temperature for life - where do we draw the line? *Trends in Microbiology*, 12(2):60–62.

Kashefi, K. and Lovley, D. R. (2003). Extending the upper temperature limit for life. *Science*, 301:943.

Kawamura, K. and Kaplan, I. R. (1987). Dicarboxylic acids generated by thermal alteration of kerogen and humic acids. *Geochimica et Cosmochimica Acta*, 51(12):3201–3207.

Kerr, R. A. (1997). Life goes to extremes in the Deep Earth - and elsewhere? *Science*, 276(5313):703–704.

Keym, M., Dieckmann, V., Horsfield, B., Erdmann, M., Galimberti, R., Kua, L. C., Leith, L., and Podlaha, O. (2006). Source rock heterogeneity of the Upper Jurassic Draupne Formation, North Viking Graben, and its relevance to petroleum generation studies. *Organic Geochemistry*, 37:220–243.

Khaddor, M., Ziyad, M., Joffre, J., and Ambles, A. (2002). Pyrolysis and characterization of the kerogen from the Moroccan Youssoufia rock phosphate. *Chemical Geology*, 186(1-2):17–30.

Khanna, R., Witt, M., Khalid Anwer, M., Agarwal, S. P., and Koch, B. P. (2008). Spectroscopic characterization of fulvic acids extracted from the rock exudate Shilajit. *Organic Geochemistry*, 39(12):1719–1724.

References

Killops, S., Funnell, R., Suggate, R., Sykes, R., Peters, K., Walters, C., Woolhouse, A., Weston, R., and Boudou, J. (1998). Predicting generation and expulsion of paraffinic oil from vitrinite-rich coals. *Organic Geochemistry*, 29(1-3):1–21.

Killops, S., Jarvie, D., Sykes, R., and Funnell, R. (2002). Maturity-related variation in the bulk-transformation kinetics of a suite of compositionally related New Zealand coals. *Marine and Petroleum Geology*, 19(9):1151–1168.

Killops, S. D. and Frewin, N. L. (1994). Triterpenoid diagenesis and cuticular preservation. *Organic Geochemistry*, 21(12):1193–1209.

Killops, S. D. and Killops, V. J. (2004). *An introduction to organic geochemistry*. Blackwell Publishing, Oxford, England.

Killops, S. D., Woolhouse, A. D., Weston, R. J., and Cook, R. A. (1994). A geochemical appraisal of oil generation in the Taranaki Basin, New Zealand. *AAPG Bulletin*, 78(10):1560–1585.

King, R. R., Naish, T. R., Browne, G. H., Field, B. D., and Edbrooke, S. W. (1999). Cretaceous-Cenozoic geology and petroleum systems of the Taranaki Basin, New Zealand. Technical report, Institute of Geological & Nuclear Sciences.

Klemme, H. D. and Ulmishek, G. F. (1991). Effective petroleum source rocks of the world; stratigraphic distribution and controlling depositional factors. *AAPG Bulletin*, 75(12):1809–1851.

Köller, C. (2002). *Paläochemotaxonomie von Torfen Nordwestdeutschlands*. PhD thesis, Carl von Ossietzky Universität Oldenburg.

Knicker, H., Scaroni, A. W., and Hatcher, P. G. (1996). ^{13}C and ^{15}N NMR spectroscopic investigation on the formation of fossil algal residues. *Organic Geochemistry*, 24(6-7):661–669.

Kolattukudy, P. E. (1966). Biosynthesis of wax in brassica oleracea. Relation of fatty acids to wax. *Biochemistry*, 5(7):2265–2275.

Kolattukudy, P. E. (1970). Plant waxes. *Lipids*, 5(2):259–275.

Kolattukudy, P. E., Croteau, R., and Buckner, J. S. (1976). Biochemistry of plant waxes. In Kolattukudy, P. E., editor, *Chemistry and Biochemistry of Natural Waxes*, pages 289–347. Elsevier, Amsterdam.

Koopmans, M. P., Schaeffer-Reiss, C., de Leeuw, J. W., Lewan, M. D., Maxwell, J. R., Schaeffer, P., and Sinninghe Damsté, J. S. (1997). Sulphur and oxygen sequestration of n-C_{37} and n-C_{38} unsaturated ketones in an immature kerogen and the release of their carbon skeletons during early stages of thermal maturation. *Geochimica et Cosmochimica Acta*, 61(12):2397–2408.

Krumholz, L. R. (2000). Microbial communities in the deep subsurface. *Hydrogeology Journal*, 8:4–10.

Krumholz, L. R., McKinley, J. P., Ulrich, G. A., and Suflita, J. M. (1997). Confined subsurface microbial communities in Cretaceous rock. *Nature*, 386:64–66.

Kuehn, D. W., Snyder, R. W., Davis, A., and Painter, P. C. (1982). Characterization of vitrinite concentrates. 1. Fourier transform infrared studies. *Fuel*, 61:682–694.

Kvenvolden, K. A. (1966). Molecular distributions of normal fatty acids and paraffins in some lower Cretaceous sediments. *Nature*, 209(5023):573–577.

Kvenvolden, K. A. (1967). Normal fatty acids in sediments. *Journal of the American Oil Chemists' Society*, 44(11):628–636.

Lafargue, E., Marquis, F., and Pillot, D. (1998). Rock-Eval 6 applications in hydrocarbon exploration and soil communication studies. *Revue de l'Institut Français du Pétrole*, 53(4):421–437.

References

Landis, E. R. and Weaver, J. N. (1993). Global coal occurrence. In Law, B. E. and Dudley, D. R., editors, *Hydrocarbons from coal.*, volume 38, pages 1–12. AAPG Bulletin.

Larter, S. R. and Douglas, A. G. (1982). Pyrolysis methods in organic geochemistry: An overview. *Journal of Analytical and Applied Pyrolysis*, 4(1):1–19.

Larter, S. R. and Horsfield, B. (1993). Determination of structural components of kerogens by the use of analytical pyrolysis. In Engel, M. H. and Macko, S. A., editors, *In Organic Geochemistry*, pages 271–287. Plenum Press.

Lehne, E. and Dieckmann, V. (2007). The significance of kinetic parameters and structural markers in source rock asphaltenes, reservoir asphaltenes and related source rock kerogens, the Duvernay Formation (WCSB). *Fuel*, 86:887–901.

Leontaris, K. J. and Mansoori, G. A. (1988). Asphaltene deposition: A survey of field experiences and research approaches. *Journal of Petroleum Science and Engineering*, 3:229–239.

Levine, J. (1993). Coalification: the evolution of coal as source rock and reservoir rock for oil and gas. In Law, B. and Dudley, D., editors, *Hydrocarbons from Coal. AAPG Studies in Geology*, volume 38, pages 39–77.

L'Haridon, S., Reysenbach, A.-L., Glénat, P., Prieur, D., and Jeanthon, C. (1995). Hot subterranean biosphere in a continental oil reservoir. *Nature*, 377:223–224.

Lin, L. H., Hall, J., Lippmann-Pipke, J., Ward, J. A., Sherwood-Lollar, B., Deflaun, M., Rothmel, R., Moser, D., Gihring, T. M., Mislowack, B., and Onstott, T. C. (2005). Radiolytic H_2 in continental crust: Nuclear power for deep subsurface microbial communities. *Geochemistry, Geophysics, Geosystems*, 6(7).

Lin, L. H., Wang, P.-L., Rumble, D., Lippmann-Pipke, J., Boice, E., Pratt, L. M., Lollar, B. S., Brodie, E. L., Hazan, T. C., Andersen, G. L., DeSantis, T. Z., Moser, D. P., and Onstott, T. C. (2006). Long-term sustainability of a high-energy, low-diverstry crustal biome. *Science*, 314:479–482.

Lis, G. P., Mastalerz, M., Schimmelmann, A., Lewan, M. D., and Stankiewicz, B. A. (2005). FTIR absorption indices for thermal maturity in comparison with vitrinite reflectance R_0 in type-II kerogens from Devonian black shales. *Organic Geochemistry*, 36(11):1533–1552.

Littke, R., Krooss, B. M., Idiz, E. F., and Frielingsdorf, J. (1995). Molecular nitrogen in natural gas accumulations; generation from sedimentary organic matter at high temperatures. *AAPG Bulletin*, 79(3):410–430.

Lovley, D. R. and Chapelle, F. H. (1995). Deep surface microbial processes. *Reviews of Geophysics*, 33(3):356–381.

Ludwig, B. and Balkenhol, R. (2001). Quantification of the acidification potential of pyrite containing sediment. *Acta Hydrochimica et Hydrobiologica*, 29(2):118–128.

Madigan, M. T. and Martinko, J. M. (2006). *Brocks - Biology Of Microorganisms*. Pearson Education, Inc., Upper Saddle River, 11 edition.

Mahlstedt, N., Horsfield, B., and Dieckmann, V. (2008). Second order reactions as a prelude to gas generation at high maturity. *Organic Geochemistry*, 39(8):1125–1129.

Mangelsdorf, K. and Rullkötter, J. (2003). Natural supply of oil-derived hydrocarbons into marine sediments along the California continental margin during the late Quaternary. *Organic Geochemistry*, 34(8):1145–1159.

Mangelsdorf, K., Zink, K., Birrien, J., and Toffin, L. (2005). A quantitative assessment of pressure dependent adaptive changes in the membrane lipids of a piezosensitive deep sub-seafloor bacterium. *Organic Geochemistry*, 36(11):1459–1479.

References

Mansuy, L., Landais, P., and Ruau, O. (1995). Importance of the reacting medium in artificial maturation of a coal by confined pyrolysis. 1. hydrocarbons and polar compounds. *Energy & Fuels*, 9:691–703.

Marchand, A. and Conard, J. (1980). Electron paramagnetic resonance in kerogen studies. In Durand, B., editor, *Kerogen. Insoluble Organic Matter from Sedimentary Rocks.*, pages 243–270. Editions Technip., Paris.

Marchand, A., Libert, P., and Combaz, A. (1969). Essai de caracterisation physico-chimique de la diagénèse de quelques roches biologiquement homogènes. *Revue de l'Institut Françcais du Pétrole*, 24:3–20.

McKinley, J. P. (2001). The use of organic geochemistry and the importance of sample scale in investigations of lithologically heterogeneous microbial ecosystems. In Fredrickson, J. K. and Fletcher, M., editors, *Subsurface Microbiology and Biogeochemistry*, pages 173–192. Wiley-Liss, Inc.

Michaelis, W. and Richnow, H. H. (1989). Structural studies of marine and riverine humic matter by chemical degradation. *The Science of The Total Environment*, 81/82:41–50.

Mitchell, D. L. and Speight, J. G. (1973). The solubility of asphaltenes in hydrocarbon solvents. *Fuel*, 52:149–152.

Morita, R. Y. and ZoBell, C. E. (1955). Occurrence of bacteria in pelagic sediments collected during the mid-Pacific expedition. *Deep Sea Research*, 3(1):66–73.

Mortimer, N., Tulloch, A. J., and Ireland, T. R. (1997). Basement geology of Taranaki and Wanganui Basins, New Zealand. *New Zealand Journal of Geology and Geophysics*, 40(2):223–236.

Mössbauer, R. L. (1962). Recoilless nuclear resonance absorption of gamma radiation: A new principle yields gamma lines of extreme narrowness for measurements of unprecedented accuracy. *Science*, 137:721–775.

Muir, R., Bradshaw, J. D., Weaver, S. D., and Laird, M. G. (2000). The influence of basement structure on the evolution of the Taranaki Basin, New Zealand. *Journal of the Geological Society, London*, 157:1179–1185.

Muir, R. J., Ireland, T. R., Weaver, S. D., and Bradshaw, J. D. (1996). Ion microprobe dating of Paleozoic granitoids: Devonian magmatism in New Zealand and correlations with Australia and Antarctica. *Chemical Geology*, 127(1-3):191–210.

Narayanan, C. R. and Lyer, K. N. (1965). Regeneration of steroid alcohols from their methyl ethers. *Journal of Organic Chemistry*, 30:1734–1736.

Nealson, K. H. and Cox, B. L. (2002). Microbial metal-ion reduction and Mars: Extraterrestrial expectations? *Current Opinion in Microbiology*, 5(3):296–300.

Nimz, H. (1974). Beech lignin - proposal of a constitutional scheme. *Angewandte Chemie International Edition*, 13(5):313–321.

Norgate, C. M., Boreham, C. J., and Wilkins, A. J. (1999). Changes in hydrocarbon maturity indices with coal rank and type, Buller Coalfield, New Zealand. *Organic Geochemistry*, 30(8):985–1010.

Oberlin, A., Boulmier, J. L., and Villey, M. (1980). Electron microscopie study of kerogen microtexture. Selected criteria for determining the evolution path and evolution stage of kerogen. In Durand, B., editor, *Kerogen, Insoluble Organic Matter from Sedimentary Rocks*, pages 191–241. Editions Technip, Paris.

Oldenburg, T. B. P., Rullkötter, J., Böttcher, M. E., and Nissenbaum, A. (2000). Molecular and isotopic characterization of organic matter in recent and sub-recent sediments from the Dead Sea. *Organic Geochemistry*, 31(4):251–265.

References

Olson, G. J., Turbak, S. C., and McFeters, G. A. (1979). Impact of western coal mining–II. Microbiological studies. *Water Research*, 13(11):1033–1041.

Onstott, T. C., Tobin, K., Dong, H., DeFlaun, M. F., Fredrickson, J. K., Bailey, T., Brockman, F. J., Kieft, T. L., Peacock, A., White, D. C., Balkwill, D., Phelps, T. J., and Boone, D. R. (1997). The deep gold mines of South Africa: windows into the subsurface biosphere. *SPIE 42nd Annual Proceedings*, 3111:344–357.

Orr, W. L. (1986). Kerogen/asphaltene/sulfur relationships in sulfur-rich Monterey oils. *Organic Geochemistry*, 10(1-3):499–516.

Otera, J. (1993). Transesterification. *Chemistry Reviews*, 93(4):1449–1470.

Painter, P. C., Snyder, R. W., Starsinic, M., Coleman, M. M., Kuehn, D. W., and Davis, A. (1981). Concerning the application of FTIR to the study of coal: A critical assessment of band assignments and the application of spectral analysis programs. *Applied Spectroscopy*, 35:475–485.

Palacas, J., Anders, D., and King, J. (1984). South florida basin - prime example of carbonate source rocks of petroleum. In Palacas, J., editor, *Petroleum Geochemistry and Source Rock Potential of Carbonates Rocks. American Associate of Petroleum Geologists Bulletin*, volume 18, pages 71–96.

Palmer, J. A. and Andrews, P. B. (1993). Cretaceous-Tertiary sedimentation and implied tectonic controls on the structural evolution of Taranaki Basin, New Zealand. In Ballance, P. F. and Hsü, K. J., editors, *South Pacific Sedimentary Basins. Sedimentary Basins of the World 2*, pages 309–328.

Panikov, N. S. and Sizova, M. V. (2007). Growth kinetics of microorganisms isolated from alaskan soil and permafrost in solid media frozen down to -35°C. *FEMS Microbiology Ecology*, 59(2):500–512.

Parkes, R. J., Cragg, B. A., Bale, S. J., Getlifff, J. M., Goodman, K., Rochelle, P. A., Fry, J. C., Weightman, A. J., and Harvey, S. M. (1994). Deep bacterial biosphere in Pacific Ocean sediments. *Nature*, 371(6496):410–413.

Parkes, R. J., Cragg, B. A., and Wellsbury, P. (2000). Recent studies on bacterial populations and processes in subseafloor sediments: A review. *Hydrogeology Journal*, 8(1):11–28.

Parkes, R. J., Wellsbury, P., Mather, I. D., Cobb, S. J., Cragg, B. A., Hornibrook, E. R. C., and Horsfield, B. (2007). Temperature activation of organic matter and minerals during burial has the potential to sustain the deep biosphere over geological timescales. *Organic Geochemistry*, 38(6):845–852.

Pedersen, K. (2000). Exploration of deep intraterrestrial microbial life: Current perspectives. *FEMS Microbiology Letters*, 185(1):9–16.

Petersen, H., Rosenberg, P., and Nytoft, H. (2008). Oxygen groups in coals and alginite-rich kerogen revisited. *International Journal of Coal Geology*, 74(2):93–113.

Petsch, S. T., Berner, R. A., and Eglinton, T. I. (2000). A field study of the chemical weathering of ancient sedimentary organic matter. *Organic Geochemistry*, 31(5):475–487.

Powell, T. (1984). Some aspects of the hydrocarbon geochemistry of a middle Devonian barrier-reef comples. In Palacas, J., editor, *Petroleum Geochemistry and Source Rock Potential of Carbonates Rocks. American Associate of Petroleum Geologists Bulletin*, volume 18, pages 45–61.

Powell, T. (1991). Petroleum source rock assessment in non-marine sequences: Pyrolysis and petrographic analysis of Australian coals and carbonaceous shales. *Organic Geochemistry*, 17(3):375–394.

References

Prahl, F. G. and Pinto, L. A. (1987). A geochemical study of long-chain n-aldehydes in washington coastal sediments. *Geochimica et Cosmochimica Acta*, 51:1573–1582.

Putschew, A., Schaeffer-Reiss, C., Schaeffer, P., Koopmans, M. P., de Leeuw, J. W., Lewan, M. D., Sinninghe Damsté, J. S., and Maxwell, J. R. (1998). Release of sulfur- and oxygen-bound components from a sulfur-rich kerogen during simulated maturation by hydrous pyrolysis. *Organic Geochemistry*, 29(8):1875–1890.

Radke, M., Willsch, H., and Welte, D. H. (1980). Preparative hydrocarbon group type determination by automated medium pressure liquid chromatography. *Analytical Chemistry*, 52(3):406–411.

Ratledge, C. and Wilkinson, S. G. (1988). *Microbial lipids*, volume 2. Academic Press, London.

Revill, A. T., Volkman, J. K., O'Leary, T., Summons, R. E., Boreham, C. J., Banks, M. R., and Denwer, K. (1994). Hydrocarbon biomarkers, thermal maturity, and depositional setting of Tasmanite oil shales from Tasmania, Australia. *Geochimica et Cosmochimica Acta*, 58(18):3803–3822.

Richnow, H. H., Jenisch, A., and Michaelis, W. (1992). Structural investigations of sulphur-rich macromolecular oil fractions and a kerogen by sequential chemical degradation. *Organic Geochemistry*, 19(4-6):351–370.

Riffaldi, R. and Schnitzer, M. (1972). Electron spin resonance spectrometry of humic substances. *Soil Science Society of America Proceedings*, 36:301–305.

Robin, P. L. (1975). *Caractérisation des kérogènes et de leur évolution par spectroscopie infrarouge*. PhD thesis, Université de Louvain.

Rothschild, L. J. (1990). Earth analogs for Martian life. Microbes in evaporites, a new model system for life on Mars. *Icarus*, 88(1):246–260.

Roussel, E. G., Bonavita, M.-A. C., Querellou, J., Cragg, B. A., Webster, G., Prieur, D., and Parkes, R. J. (2008). Extending the sub-sea-floor biosphere. *Science*, 320:1046.

Rütters, H., Sass, H., Cypionka, H., and Rullkötter, J. (2002). Phospholipid analysis as a tool to study complex microbial communities in marine sediments. *Journal of Microbiological Methods*, 48:149–160.

Rubinsztain, Y., Ioselis, P., Ikan, R., and Aizenshtat, Z. (1984). Investigations on the structural units of melanoidins. *Organic Geochemistry*, 6:791–804.

Rueter, P., Rabus, R., Wilkes, H., Aeckersberg, F., Rainey, F. A., Jannasch, H. W., and Widdel, F. (1994). Anaerobic oxidation of hydrocarbons in crude oil by new types of sulphate-reducing bacteria. *Nature*, 372:455–458.

Rullkötter, J. and Michaelis, W. (1990). The structure of kerogen and related materials. A review of recent progress and future trends. *Organic Geochemistry*, 16(4-6):829–852.

Russell, N. J. and Fukunaga, N. (1990). A comparison of thermal adaptation of membrane lipids in psychrophilic and thermophilic bacteria. *FEMS Microbiological Reviews*, 75:71–182.

Sainsbury, M. (1970). Friedelin and epifriedelinol from the bark of *Prunus turfosa* and a review of their natural distribution. *Phytochemistry*, 9(10):2209–2215.

Sakai, A. and Wardle, P. (1978). Freezing resistance of New Zealand trees and shrubs. *New Zealand Journal of Ecology*, 1:51–61.

Sansone, F. J. and Martens, C. S. (1981). Methane production from acetate and associated methane fluxes from anoxic coastal sediments. *Science*, 211:707–709.

Sansone, F. J. and Martens, C. S. (1982). Volatile fatty acid cycling in organic-rich marine sediments. *Geochimica et Cosmochimica Acta*, 46(9):1575–1589.

References

Santosh, M. and Omori, S. (2008a). CO_2 flushing: A plate tectonic perspective. *Gondwana Research*, 13(1):86–102.

Santosh, M. and Omori, S. (2008b). CO_2 windows from mantle to atmosphere: Models on ultrahigh-temperature metamorphism and speculations on the link with melting of snowball Earth. *Gondwana Research*, 14(1-2):82–96.

Sarret, G., Connan, J., Kasrai, M., Bancroft, G. M., Charrié-Duhaut, A., Lemoine, S., Adam, P., Albrecht, P., and Eybert-Bérard, L. (1999). Chemical forms of sulfur in geological and archeological asphaltenes from Middle East, France, and Spain determined by sulfur K- and L-edge X-ray absorption near-edge structure spectroscopy. *Geochimica et Cosmochimica Acta*, 63(22):3767–3779.

Schaefer, J. and Stejskal, E. O. (1976). Carbon-13 nuclear magnetic resonance of polymers spinning at the magic angle. *Journal of the American Chemical Society*, 98(4):1031–1032.

Schaeffer, P., Harrison, W. N., Keely, B. J., and Maxwell, J. R. (1995). Product distributions from chemical degradation of kerogens from a marl from a Miocene evaporitic sequence (Vena del Gesso, N. Italy). *Organic Geochemistry*, 23(6):541–554.

Schaeffer-Reiss, C., Schaeffer, P., Putschew, A., and Maxwell, J. R. (1998). Stepwise chemical degradation of immature S-rich kerogens from Vena del Gesso (Italy). *Organic Geochemistry*, 29(8):1857–1873.

Schenk, H. J., Horsfield, B., Kroos, B., Schaeffer, R. G., and Schwochau, K. (1997). Kinetics of petroleum formation and cracking. In Welte, D. H., Horsfield, B., and Backer, D. R., editors, *Petroleum and Basin Evolution. Insights from Petroleum Geochemistry, Geology and Basin Modelling*, pages 231–270. Springer, Berlin.

Schnitzer, M. (1985). Nature of nitrogen in humic substances. In Mc Knight, D., Aiken, G., Wershaw, R., and MacCarthy, P., editors, *Humic Substances* in *Soil, Sediment and Water: Geochemistry, Isolation and Characterisation.*, pages 303–325. Wiley & Sons, New York & Chinchester.

Schopf, J. M. (1956). A definition of coal. *Economic Geology*, 51(6):521–527.

Schouten, S., Pavlovic, D., Sinninghe-Damsté, J. S., and de Leeuw, J. W. (1993). Nickel boride: An improved desulphurizing agent for sulphur-rich geomacromolecules in polar and asphaltene fractions. *Organic Geochemistry*, 20(7):901–909.

Schuchardt, U., Sercheli, R., and Vargas, R. M. (1998). Transesterification of vegetable oils: A review. *Journal of the Brazilian Chemical Society*, 9(1):199–210.

Schwarzbauer, J., Ricking, M., and Littke, R. (2003). Quantitation of nonextractable anthropogenic sediments after chemical degradation. *Acta Hydrochimica et Hydrobiologica*, 31(6):469–481.

Seifert, K. (1975). Carboxylic acids in petroleum and sediments. *Fortschritte der Chemie organischer Naturstoffe*, 32:1–49.

Shameel, M. (1990). Physicochemical studies and fatty acids from certain seaweeds. *Botanica Marina*, 33:429–432.

Sherwood-Lollar, B., Couloume, L. G., Slater, G. F., Ward, J., Moser, D. P., Gihring, T. M., Lin, L. H., and Onstott, T. C. (2006). Unravelling abiogenic and biogenic sources of methane in the Earth's deep subsurface. *Chemical Geology*, 226(3-4):328-339.

Simoneit, B., Leif, R., and Ishiwatari, R. (1996). Phenols in hydrothermal petroleums and sediment bitumen from Guaymas Basin, Gulf of California. *Organic Geochemistry*, 24:377–388.

Sinninghe Damsté, J. S. and De Leeuw, J. W. (1990). Analysis, structure and geochemical significance of organically-bound sulphur in the geosphere: State of the art and future research. *Organic Geochemistry*, 16(4-6):1077–1101.

References

Sinninghe Damsté, J. S., Keely, B. J., Betts, S. E., Baas, M., Maxwell, J. R., and de Leeuw, J. W. (1993). Variations in abundances and distributions of isoprenoid chromans and long-chain alkylbenzenes in sediments of the Mulhouse Basin: a molecular sedimentary record of palaeosalinity. *Organic Geochemistry*, 20:1201–1215.

Sinninghe Damsté, J. S., Rijpstra, W. I. C., De Leeuw, J. W., and Schenck, P. A. (1989a). The occurrence and identification of series of organic sulphur compounds in oils and sediment extracts: II. Their presence in samples from hypersaline and non-hypersaline palaeoenvironments and possible application as source, palaeoenvironmental and maturity indicators. *Geochimica et Cosmochimica Acta*, 53(6):1323–1341.

Sinninghe Damsté, J. S., Schouten, S., de Leeuw, J. W., van Duin, A. C. T., and Geenevasen, J. A. J. (1999). Identification of novel sulfur-containing steroids in sediments and petroleum: Probable incorporation of sulfur into [delta]5,7-sterols during early diagenesis. *Geochimica et Cosmochimica Acta*, 63(1):31–38.

Sinninghe Damsté, J. S., van Koert, E. R., Kock-van Dalen, A. C., de Leeuw, J. W., and Schenck, P. A. (1989b). Characterisation of highly branched isoprenoid thiophenes occurring in sediments and immature crude oils. *Organic Geochemistry*, 14(5):555–567.

Sinninghe-Damsté, J. S., Eglinton, T. I., De Leeuw, J. W., and Schenck, P. A. (1989). Organic sulphur in macromolecular sedimentary organic matter: I. Structure and origin of sulphur-containing moieties in kerogen, asphaltenes and coal as revealed by flash pyrolysis. *Geochimica et Cosmochimica Acta*, 53(4):873–889.

Siskin, M. and Katritzky, A. R. (1991). Reactivity of organic compounds in hot water: Geochemical and technological implications. *Science*, 254(5029):231–237.

Sogin, M. L., Morrison, H. G., Huber, J. A., Welch, D. M., Huse, S. M., Neal, P. R., Arrieta, J. M., and Herndl, G. J. (2006). Microbial diversity in the deep sea and the underexplored "rare biosphere". *Proceedings of the National Academy of Sciences, USA*, 103(32):12115–12120.

Speight, J. G., Long, R. B., and Trowbridge, T. D. (1984). Factors influencing the separation of asphaltenes from heavy petroleum feedstocks. *Fuel*, 63:616–620.

Sørensen, J., Christensen, D., and Jørgensen, B. B. (1981). Volatile fatty acids and hydrogen as substrates for sulfate-reducing bacteria in anaerobic marine sediments. *Applied And Environmental Microbiology*, 42(1):5–11.

Staley, J. T. (1999). Bacterial biodiversity: A time for place. *ASM News*, 65:681–187.

Stetter, K. O., Huber, R., Blöchl, E., Kurr, M., Eden, R. D., Fielder, M., Cash, H., and Vance, I. (1993). Hyperthemophilic archaea are thriving in deep North Sea and Alaskan oil reservoirs. *Nature*, 365:743–745.

Stevens, K. O. and McKinley, J. P. (1995). Litoautotrophic microbial ecosystems in deep basalt aquifers. *Science*, 270:450–454.

Stock, L. M. and Tse, K.-T. (1983). Ruthenium tetroxide catalysed oxidation of Illinois no.6 coal and some representative hydrocarbons. *Fuel*, 62:974–976.

Stock, L. M. and Wang, S.-H. (1986). Ruthenium tetraoxide catalysed oxidation of Illinois no.6 coal. the formation of aliphatic and benzene carboxylic acids. *Fuel*, 65:1552–1562.

Stokes, G. G. (1901). *Mathematical and Physical Papers*, volume III. University Press, Cambridge.

Stopes, M. (1935). On the petrology of banded bituminous coals. *Fuel*, 14:4–13.

References

Stout, S. A., Boon, J. J., and Spackman, W. (1988). Molecular aspects of the peatification and early coalification of angiosperm and gymnosperm woods. *Geochimica et Cosmochimica Acta*, 52(2):405–414.

Strapoc, D., Picardal, F. W., Turich, C., Schaperdoth, I., Macalady, J. L., Lipp, J. S., Lin, Y.-S., Ertefai, T. F., Schubotz, F., Hinrichs, K.-U., Mastalerz, M., and Schimmelmann, A. (2008). Methane-producing microbial community in a coal bed of the Illinois Basin. *Applied and Environmental Microbiology*, 74(8):2424–2432.

Sturt, H. F., Summons, R. E., Smith, K., Elvert, M., and Hinrichs, K.-U. (2004). Intact polar membrane lipids in prokaryotes and sediments deciphered by high-performance liquid chromatography/electrospray ionization multistage mass spectrometry - new biomarkers for biogeochemistry and microbial ecology. *Rapid Communications in Mass Spectrometry*, 18:617–628.

Suggate, R. P. (2000). The rank (Sr) scale: Its basis and its applicability as a maturity index for all coals. *New Zealand Journal of Geology and Geophysics*, 43:521–553.

Suggate, R. P. and Dickinson, W. W. (2004). Carbon NMR of coals: The effects of coal type and rank. *International Journal of Coal Geology*, 57(1):1–22.

Sweeney, J. J. and Burnham, A. K. (1990). Evaluation of a simple model of vitrinite reflectance based on chemical kinetics. *The American Association of Petroleum Geologists Bulletin*, 74(10):1559–1570.

Sykes, R. and Johansen, P. E. (2007). Maturation characteristics of the New Zealand Coal Band: Part 1 - Evolution of oil and gas products. *The 23rd International Meeting on Organic Geochemistry, Book of Abstracts*, page 571.

Sykes, R. and Snowdon, L. R. (2002). Guidelines for assessing the petroleum potential of coaly source rocks using Rock-Eval pyrolysis. *Organic Geochemistry*, 33(12):1441–1455.

Takeda, N. and Asakawa, T. (1988). Study of petroleum generation by pyrolysis–I. Pyrolysis experiments by Rock-Eval and assumption of molecular structural change of kerogen using ^{13}C-NMR. *Applied Geochemistry*, 3(5):441–453.

Tegelaar, E. W., de Leeuw, J. W., Derenne, S., and Largeau, C. (1989). A reappraisal of kerogen formation. *Geochimica et Cosmochimica Acta*, 53(11):3103–3106.

Theuerkorn, K., Horsfield, B., Wilkes, H., di Primio, R., and Lehne, E. (2008). A reproducible and linear method for separating asphaltenes from crude oil. *Organic Geochemistry*, 39(8):929–934.

Tissot, B., Durand, B., Espitalie, J., and Combaz, A. (1974). Influence of nature and diagenesis of organic matter in formation of petroleum. *AAPG Bulletin*, 58(3):499–506.

Tissot, B. T. and Welte, D. H. (1984a). Diagenesis, catagenesis and metagenesis of kerogen. In Tissot, B. T. and Welte, D. H., editors, *Petroleum Formation and Occurrence*, pages 160–169. Springer, Heidelberg.

Tissot, B. T. and Welte, D. H. (1984b). *Petroleum Formation and Occurrence*. Springer, Heidelberg.

Trager, E. A. (1924). Kerogen and its relation to the origin of oil. *American Association of Petroleum Geologists Bulletin*, 8:301–311.

van Krevelen, D. W. (1961). *Coal: Typology - Chemistry - Physics - Constitution*. Elsevier, The Netherlands, 1st edition.

Vandenbroucke, M. and Largeau, C. (2007). Kerogen origin, evolution and structure. *Organic Geochemistry*, 38(5):719–833.

Vassoevitch, N. B., Korchaginam, Y. I., Lopation, N. V., and Chernyshev, V. V. (1969). Principal phase of oil formation. *International Geology Review*, 12:1276–1296.

References

Vieth, A., Mangelsdorf, K., Sykes, R., and Horsfield, B. (2008). Water extraction of coals - potential for estimating low molecular weight organic acids as carbon feedstock for the deep terrestrial biosphere. *Organic Geochemistry*, 39(8):985–991.

Vu, T. T. A. (2008). *Origin and Maturation of Organic Matter in New Zealand Coals*. PhD thesis, Ernst-Moritz-Arndt-University Greifswald.

Vu, T. T. A., Horsfield, B., and Sykes, R. (2008). Influence of in-situ bitumen on the generation of gas and oil in new zealand coals. *Organic Geochemistry*, 39(11):1606–1619.

Vu, T. T. A., Zink, K. G., Mangelsdorf, K., Sykes, R., Wilkes, H., and Horsfield, B. (2009). Changes in bulk properties and molecular compositions within New Zealand Coal Band solvent extracts from early diagenetic to catagenetic maturity levels. *Organic Geochemistry*, 40(9):963–977.

W. Hogg, R. and T. Gillan, F. (1984). Fatty acids, sterols and hydrocarbons in the leaves from eleven species of mangrove. *Phytochemistry*, 23(1):93–97.

Wächtershäuser, G. (1990). Evolution of the first metabolic cycles. *Proceedings of the National Academy of Sciences of the United States of America*, 87(1):200–204.

Wang, S. and Rullkötter, J. (1997). Neutral polar lipids in freshwater and alkaline lacustrine sediments of the Green River Formation, Wyoming, U.S.A. In *Abstracts of the 18th International Meeting on Organic Geochemistry, 22.-26.09, Maastricht, The Netherlands*, pages 255–256.

Watanabe, A., McPhail, D. B., Maie, N., Kawasaki, S., Anderson, H. A., and Cheshire, M. V. (2005). Electron spin resonance characteristics of humic acids from a wide range of soil types. *Organic Geochemistry*, 36(7):981–990.

Weijers, J. W. H., Schouten, S., Spaargaren, O. C., and Sinninghe Damsté, J. S. (2006). Occurrence and distribution of tetraether membrane lipids in soils: Implications for the use of the TEX86 proxy and the BIT index. *Organic Geochemistry*, 37(12):1680–1693.

Wellsbury, P., Goodman, K., Barth, T., Cragg, B. A., Barnes, S. P., and Parkes, J. R. (1997). Deep marine biosphere fuelled by increasing organic matter availability during burial and heating. *Nature*, 388:573–576.

Werner-Zwanziger, U., Lis, G., Mastalerz, M., and Schimmelmann, A. (2005). Thermal maturity of type II kerogen from the New Albany Shale assessed by ^{13}C CP/MAS NMR. *Solid State Nuclear Magnetic Resonance*, 27(1-2):140–148.

White, D. C., Davis, W. M., Nickels, J. S., King, J. D., and Bobbie, R. J. (1979). Determination of the sedimentary microbial biomass by extractible lipid phosphate. *Oecologia*, 40:51–62.

Whiteman, W. B., Coleman, D. C., and Wiebe, W. J. (1998). Prokariontes: The unseen majority. *Proceedings of the National Academy of Sciences, USA*, 95:6578–6583.

Wilhelms, A., Larter, S. R., Head, I., Farrimond, P., di Primio, R., and Zwach, C. (2001). Biodegradation of oil in uplifted basins prevented by deep-burial sterilization. *Nature*, 411(6841):1034–1037.

Wilkes, H., Disko, U., and Horsfield, B. (1998). Aromatic aldehydes and ketones in the Posidonia Shale, Hils Syncline, Germany. *Organic Geochemistry*, 29(1-3):107–117.

Willsch, H., Clegg, H., Horsfield, B., Radke, M., and Wilkes, H. (1997). Liquid chromatographic separation of sediment, rock, and coal extracts and crude oil into compound classes. *Analytical Chemistry*, 69(20):4203–4209.

Wilson, M. A. and Hatcher, P. G. (1988). Detection of tannins in modern and fossil barks and in plant residues by high-resolution solid-state ^{13}C nuclear magnetic resonance. *Organic Geochemistry*, 12(6):539–546.

References

Wollenweber, J., Schwarzbauer, J., Littke, R., Wilkes, H., Armstroff, A., and Horsfield, B. (2006). Characterisation of non-extractable macromolecular organic matter in Palaeozoic coals. *Palaeogeography, Palaeoclimatology, Palaeoecology*, 240(1-2):275–304.

Yabuta, H., Fukushima, M., Kawasaki, M., Tanaka, F., Kobayashi, T., and Tatsumi, K. (2008). Multiple polar components in poorly-humified humic acids stabilizing free radicals: Carboxyl and nitrogen-containing carbons. *Organic Geochemistry*, 39(9):1319–1335.

Yamamoto, M., Okino, T., Sugisaki, S., and Sakamoto, T. (2008). Late Pleistocene changes in terrestrial biomarkers in sediments from the central Arctic Ocean. *Organic Geochemistry*, 39(6):754–763.

Yano, Y., Nakayama, A., Ishihara, K., and Saito, H. (1998). Adaptive changes in membrane lipids of barophilic bacteria in response to changes in growth pressure. *Applied and Environmental Microbiology*, 64:479–485.

Yu, D. M., Ranathunge, K., Huang, H., Pei, Z. Y., Franke, R., Schreiber, L., and He, C. Z. (2008). Wax crystal-sparse leaf encodes a beta-ketoacyl CoA synthase involved in biosynthesis of cuticular waxes on rice leaf. *Planta*, 228(4):675–685.

Zink, K. G., Mangelsdorf, K., Granina, L., and Horsfield, B. (2008). Estimation of bacterial biomass in subsurface sediments by quantifying intact membrane phospholipids. *Analytical and Bioanalytical Chemistry*, 390:885–896.

Zink, K.-G., Wilkes, H., Disko, U., Elvert, M., and Horsfield, B. (2003). Intact phospholipids-microbial "life markers" in marine deep subsurface sediments. *Organic Geochemistry*, 34(6):755–769.

ZoBell, C. E. (1945). The role of bacteria in the formation and transformation of petroleum hydrocarbons. *Science*, 102:364–369.

Appendix

A. Tables

A. Tables

G004541; Formic acid; M: 46 g/mol, T: 90°C = 363.15 K					
t [s]	86400	172800	259200	345600	604800
bound formic acid (ester$_0$) [mg/gTOC]	17.51	17.51	17.51	17.51	17.51
hydrol. formic acid [mg/gTOC]	2.80	3.58	3.85	4.28	5.34
resid. formic acid (ester) [mg/gTOC]	14.71	13.93	13.66	13.23	12.17
resid. formic acid (ester) [mmol/gTOC]	0.320	0.303	0.297	0.288	0.265
ln(ester/ester$_0$)	-0.18	-0.23	-0.25	-0.28	-0.36
G004541; Acetic acid; M: 60 g/mol, T: 90°C = 363.15 K					
bound acetic acid (ester$_0$) [mg/gTOC]	3.94	3.94	3.94	3.94	3.94
hydrol. acetic acid [mg/gTOC]	1.62	2.05	2.11	2.86	3.41
resid. acetic acid (ester) [mg/gTOC]	2.32	1.89	1.83	1.08	0.53
resid. acetic acid (ester) [mmol/gTOC]	0.039	0.032	0.031	0.018	0.009
ln(ester/ester$_0$)	-0.53	-0.72	-0.76	-1.30	-1.99

Table A.1.: *Hydrolysis of formic and acetic acid esters (90° C, pH 3) for 1, 2, 3, 4 and 7 d of sample G004541.*

G001980; Formic acid; M: 46 g/mol, T: 90°C = 363.15 K					
t [s]	86400	172800	259200	345600	604800
bound formic acid (ester$_0$) [mg/gTOC]	7.36	7.36	7.36	7.36	7.36
hydrol. formic acid [mg/gTOC]	0.96	0.65	1.56	1.68	2.48
resid. formic acid (ester) [mg/gTOC]	6.4	6.71	5.8	5.68	4.88
resid. formic acid (ester) [mmol/gTOC]	0.139	0.146	0.126	0.123	0.106
ln(ester/ester$_0$)	-0.14	-0.09	-0.24	-0.26	-0.41
G001980; Acetic acid; M: 60 g/mol, T: 90°C = 363.15 K					
bound acetic acid (ester$_0$) [mg/gTOC]	2.24	2.24	2.24	2.24	2.24
hydrol. acetic acid [mg/gTOC]	0.73	0.45	0.97	1.46	1.89
resid. acetic acid (ester) [mg/gTOC]	1.51	1.79	1.27	0.78	0.35
resid. acetic acid (ester) [mmol/gTOC]	0.025	0.030	0.021	0.013	0.006
ln(ester/ester$_0$)	-0.39	-0.21	-0.57	-1.05	-1.82

Table A.2.: *Hydrolysis of formic and acetic acid esters (90° C, pH 3) for 1, 2, 3, 4 and 7 d of sample G001980.*

G001996; Formic acid; M: 46 g/mol, T: 90°C = 363.15 K					
t [s]	86400	172800	259200	345600	604800
bound formic acid (ester$_0$) [mg/gTOC]	7	7	7	7	7
hydrol. formic acid [mg/gTOC]	0.7	0.75	0.77	1	0.98
resid. formic acid (ester) [mg/gTOC]	6.3	6.25	6.23	6	6.02
resid. formic acid (ester) [mmol/gTOC]	0.137	0.136	0.135	0.130	0.131
ln(ester/ester$_0$)	-0.10	-0.11	-0.12	-0.15	-0.15
G001996; Acetic acid; M: 60 g/mol, T: 90°C = 363.15 K					
bound acetic acid (ester$_0$) [mg/gTOC]	1.63	1.63	1.63	1.63	1.63
hydrol. acetic acid [mg/gTOC]	0.60	0.64	0.72	1.21	1.41
resid. acetic acid (ester) [mg/gTOC]	1.03	0.99	0.91	0.42	0.22
resid. acetic acid (ester) [mmol/gTOC]	0.017	0.017	0.015	0.007	0.004
ln(ester/ester$_0$)	-0.46	-0.46	-0.59	-1.35	-1.91

Table A.3.: *Hydrolysis of formic and acetic acid esters (90° C, pH 3) for 1, 2, 3, 4 and 7 d of sample G001996.*

G001989; Formic acid; M: 46 g/mol, T: 90°C = 363.15 K					
t [s]	86400	172800	259200	345600	604800
bound formic acid (ester$_0$) [mg/gTOC]	3.49	3.49	3.49	3.49	3.49
hydrol. formic acid [mg/gTOC]	0.24	0.32	0.23	0.2	0.281
resid. formic acid (ester) [mg/gTOC]	3.25	3.17	3.26	3.29	3.21
resid. formic acid (ester) [mmol/gTOC]	0.071	0.069	0.071	0.072	0.070
ln(ester/ester$_0$)	-0.07	-0.10	-0.07	-0.05	-0.08
G001989; Acetic acid; M: 60 g/mol, T: 90°C = 363.15 K					
bound acetic acid (ester$_0$) [mg/gTOC]	0.89	0.89	0.89	0.89	0.89
hydrol. acetic acid [mg/gTOC]	0.29	0.53	0.41	0.58	0.73
resid. acetic acid (ester) [mg/gTOC]	0.60	0.36	0.48	0.31	0.16
resid. acetic acid (ester) [mmol/gTOC]	0.010	0.006	0.008	0.005	0.003
ln(ester/ester$_0$)	-0.41	-0.92	-0.63	-1.10	-1.61

Table A.4.: *Hydrolysis of formic and acetic acid esters (90° C, pH 3) for 1, 2, 3, 4 and 7 d of sample G001989.*

A. Tables

G004541; Formic acid; M: 46 g/mol, T: 75°C = 348.15 K					
t [s]	86400	172800	259200	345600	604800
bound formic acid (ester$_0$) [mg/gTOC]	17.51	17.51	17.51	17.51	17.51
hydrol. formic acid [mg/gTOC]	2.65	3.21	3.59	3.67	4.26
resid. formic acid (ester) [mg/gTOC]	14.86	14.3	13.92	13.84	13.25
resid. formic acid (ester) [mmol/gTOC]	0.323	0.311	0.303	0.301	0.288
ln(ester/ester$_0$)	-0.17	-0.20	-0.23	-0.24	-0.28
G004541; Acetic acid; M: 60 g/mol, T: 75°C = 348.15 K					
bound acetic acid (ester$_0$) [mg/gTOC]	3.94	3.94	3.94	3.94	3.94
hydrol. acetic acid [mg/gTOC]	1.66	2.01	2.26	2.23	2.69
resid. acetic acid (ester) [mg/gTOC]	2.28	1.93	1.68	1.71	1.25
resid. acetic acid (ester) [mmol/gTOC]	0.038	0.032	0.028	0.029	0.021
ln(ester/ester$_0$)	-0.55	-0.72	-0.86	-0.82	-1.15

Table A.5.: *Hydrolysis of formic and acetic acid esters (75° C, pH 3) for 1, 2, 3, 4 and 7 d of sample G004541.*

G001980; Formic acid; M: 46 g/mol, T: 75°C = 348.15 K					
t [s]	86400	172800	259200	345600	604800
bound formic acid (ester$_0$) [mg/gTOC]	7.36	7.36	7.36	7.36	7.36
hydrol. formic acid [mg/gTOC]	0.75	1.13	1.13	1.39	1.96
resid. formic acid (ester) [mg/gTOC]	6.61	6.23	6.23	5.97	5.46
resid. formic acid (ester) [mmol/gTOC]	0.144	0.135	0.135	0.130	0.119
ln(ester/ester$_0$)	-0.11	-0.17	-0.17	-0.21	-0.30
G001980; Acetic acid; M: 60 g/mol, T: 75°C = 348.15 K					
bound acetic acid (ester$_0$) [mg/gTOC]	2.24	2.24	2.24	2.24	2.24
hydrol. acetic acid [mg/gTOC]	0.65	0.90	0.94	1.10	1.40
resid. acetic acid (ester) [mg/gTOC]	1.59	1.34	1.30	1.14	0.84
resid. acetic acid (ester) [mmol/gTOC]	0.027	0.022	0.022	0.019	0.014
ln(ester/ester$_0$)	-0.32	-0.52	-0.52	-0.67	-0.97

Table A.6.: *Hydrolysis of formic and acetic acid esters (75° C, pH 3) for 1, 2, 3, 4 and 7 d of sample G001980.*

G001996; Formic acid; M: 46 g/mol, T: 75°C = 348.15 K					
t [s]	86400	172800	259200	345600	604800
bound formic acid (ester$_0$) [mg/gTOC]	7	7	7	7	7
hydrol. formic acid [mg/gTOC]	0.51	0.67	0.69	0.43	0.82
resid. formic acid (ester) [mg/gTOC]	6.49	6.33	6.31	6.57	6.18
resid. formic acid (ester) [mmol/gTOC]	0.141	0.138	0.137	0.143	0.134
ln(ester/ester$_0$)	-0.07	-0.10	-0.10	(-0.06)	-0.12
G001996; Acetic acid; M: 60 g/mol, T: 75°C = 348.15 K					
bound acetic acid (ester$_0$) [mg/gTOC]	1.63	1.63	1.63	1.63	1.63
hydrol. acetic acid [mg/gTOC]	0.51	0.64	0.79	0.48	0.96
resid. acetic acid (ester) [mg/gTOC]	1.12	0.99	0.84	1.15	0.67
resid. acetic acid (ester) [mmol/gTOC]	0.019	0.017	0.014	0.019	0.011
ln(ester/ester$_0$)	-0.35	-0.46	-0.66	(-0.35)	-0.90

Table A.7.: *Hydrolysis of formic and acetic acid esters (75°C, pH 3) for 1, 2, 3, 4 and 7 d of sample G001996.*

G001989; Formic acid; M: 46 g/mol, T: 75°C = 348.15 K					
t [s]	86400	172800	259200	345600	604800
bound formic acid (ester$_0$) [mg/gTOC]	3.49	3.49	3.49	3.49	3.49
hydrol. formic acid [mg/gTOC]	0.25	0.25	0.3	0.23	0.23
resid. formic acid (ester) [mg/gTOC]	3.24	3.24	3.19	3.26	3.26
resid. formic acid (ester) [mmol/gTOC]	0.070	0.070	0.069	0.071	0.071
ln(ester/ester$_0$)	-0.08	-0.08	-0.10	-0.07	(-0.07)
G001989; Acetic acid; M: 60 g/mol, T: 75°C = 348.15 K					
bound acetic acid (ester$_0$) [mg/gTOC]	0.89	0.89	0.89	0.89	0.89
hydrol. acetic acid [mg/gTOC]	0.28	0.33	0.43	0.41	0.64
resid. acetic acid (ester) [mg/gTOC]	0.61	0.56	0.46	0.48	0.25
resid. acetic acid (ester) [mmol/gTOC]	0.010	0.009	0.008	0.008	0.004
ln(ester/ester$_0$)	-0.41	-0.51	-0.63	-0.63	-1.32

Table A.8.: *Hydrolysis of formic and acetic acid esters (75°C, pH 3) for 1, 2, 3, 4 and 7 d of sample G001989.*

A. Tables

G004541; Formic acid; M: 46 g/mol, T: 60°C = 333.15 K					
t [s]	86400	172800	259200	345600	604800
bound formic acid (ester$_0$) [mg/gTOC]	17.51	17.51	17.51	17.51	17.51
hydrol. formic acid [mg/gTOC]	1.39	2	1.41	2.03	3.16
resid. formic acid (ester) [mg/gTOC]	16.12	15.51	16.1	15.48	14.35
resid. formic acid (ester) [mmol/gTOC]	0.350	0.337	0.350	0.337	0.312
ln(ester/ester$_0$)	-0.08	-0.12	-0.08	-0.12	-0.20
G004541; Acetic acid; M: 60 g/mol, T: 60°C = 333.15 K					
bound acetic acid (ester$_0$) [mg/gTOC]	3.94	3.94	3.94	3.94	3.94
hydrol. acetic acid [mg/gTOC]	1.00	1.28	1.02	1.31	1.92
resid. acetic acid (ester) [mg/gTOC]	2.94	2.66	2.92	2.63	2.02
resid. acetic acid (ester) [mmol/gTOC]	0.049	0.044	0.049	0.044	0.034
ln(ester/ester$_0$)	-0.30	-0.41	-0.30	-0.41	-0.66

Table A.9.: *Hydrolysis of formic and acetic acid esters (60° C, pH 3) for 1, 2, 3, 4 and 7 d of sample G004541.*

G001980; Formic acid; M: 46 g/mol, T: 60°C = 333.15 K					
t [s]	86400	172800	259200	345600	604800
bound formic acid (ester$_0$) [mg/gTOC]	7.36	7.36	7.36	7.36	7.36
hydrol. formic acid [mg/gTOC]	0.48	no data	0.5	no data	no data
resid. formic acid (ester) [mg/gTOC]	6.88	no data	6.86	no data	no data
resid. formic acid (ester) [mmol/gTOC]	0.150	no data	0.149	no data	no data
ln(ester/ester$_0$)	-0.07	no data	-0.07	no data	no data
G001980; Acetic acid; M: 60 g/mol, T: 60°C = 333.15 K					
bound acetic acid (ester$_0$) [mg/gTOC]	2.24	2.24	2.24	2.24	2.24
hydrol. acetic acid [mg/gTOC]	0.47	no data	0.48	no data	no data
resid. acetic acid (ester) [mg/gTOC]	1.77	no data	1.76	no data	no data
resid. acetic acid (ester) [mmol/gTOC]	0.030	no data	0.029	no data	no data
ln(ester/ester$_0$)	-0.21	no data	-0.24	no data	no data

Table A.10.: *Hydrolysis of formic and acetic acid esters (60° C, pH 3) for 1, 2, 3, 4 and 7 d of sample G001980.*

G001996; Formic acid; M: 46 g/mol, T: 60°C = 333.15 K					
t [s]	86400	172800	259200	345600	604800
bound formic acid (ester$_0$) [mg/gTOC]	7	7	7	7	7
hydrol. formic acid [mg/gTOC]	0.29	0.41	0.31	0.43	0.63
resid. formic acid (ester) [mg/gTOC]	6.71	6.59	6.69	6.57	6.371
resid. formic acid (ester) [mmol/gTOC]	0.146	0.143	0.145	0.143	0.1384
ln(ester/ester$_0$)	-0.04	-0.06	-0.04	-0.06	-0.09
G001996; Acetic acid; M: 60 g/mol, T: 60°C = 333.15 K					
bound acetic acid (ester$_0$) [mg/gTOC]	1.63	1.63	1.63	1.63	1.63
hydrol. acetic acid [mg/gTOC]	0.30	0.40	0.30	0.41	0.61
resid. acetic acid (ester) [mg/gTOC]	1.33	1.23	1.33	1.22	1.020
resid. acetic acid (ester) [mmol/gTOC]	0.022	0.021	0.022	0.020	0.017
ln(ester/ester$_0$)	-0.20	-0.25	-0.20	-0.30	-0.46

Table A.11.: *Hydrolysis of formic and acetic acid esters (60° C, pH 3) for 1, 2, 3, 4 and 7 d of sample G001996.*

G001989; Formic acid; M: 46 g/mol, T: 60°C = 333.15 K					
t [s]	86400	172800	259200	345600	604800
bound formic acid (ester$_0$) [mg/gTOC]	3.49	3.49	3.49	3.49	3.49
hydrol. formic acid [mg/gTOC]	0.17	0.19	0.17	0.19	0.22
resid. formic acid (ester) [mg/gTOC]	3.32	3.3	3.32	3.3	3.27
resid. formic acid (ester) [mmol/gTOC]	0.072	0.072	0.072	0.072	0.071
ln(ester/ester$_0$)	-0.05	-0.05	-0.05	-0.05	-0.076
G001989; Acetic acid; M: 60 g/mol, T: 60°C = 333.15 K					
bound acetic acid (ester$_0$) [mg/gTOC]	0.89	0.89	0.89	0.89	0.89
hydrol. acetic acid [mg/gTOC]	0.21	0.22	0.20	0.22	0.31
resid. acetic acid (ester) [mg/gTOC]	0.68	0.67	0.69	0.67	0.58
resid. acetic acid (ester) [mmol/gTOC]	0.011	0.011	0.012	0.011	0.010
ln(ester/ester$_0$)	-0.31	-0.31	-0.22	-0.31	-0.41

Table A.12.: *Hydrolysis of formic and acetic acid esters (60° C, pH 3) for 1, 2, 3, 4 and 7 d of sample G001989.*

A. Tables

G004541; Formic acid; M: 46 g/mol, T: 45°C = 318.15 K					
t [s]	86400	172800	259200	345600	604800
bound formic acid (ester$_0$) [mg/gTOC]	17.51	17.51	17.51	17.51	17.51
hydrol. formic acid [mg/gTOC]	0.87	0.53	0.7	1.4	1.1
resid. formic acid (ester) [mg/gTOC]	16.64	16.98	16.81	16.11	15.94
resid. formic acid (ester) [mmol/gTOC]	0.362	0.369	0.365	0.350	0.347
ln(ester/ester$_0$)	-0.05	-0.03	-0.04	-0.08	-0.09
G004541; Acetic acid; M: 60 g/mol, T: 45°C = 318.15 K					
bound acetic acid (ester$_0$) [mg/gTOC]	3.94	3.94	3.94	3.94	3.94
hydrol. acetic acid [mg/gTOC]	0.76	0.39	0.48	1.22	1.01
resid. acetic acid (ester) [mg/gTOC]	3.18	3.55	3.46	2.72	2.93
resid. acetic acid (ester) [mmol/gTOC]	0.053	0.059	0.058	0.045	0.049
ln(ester/ester$_0$)	-0.22	-0.11	-0.13	-0.38	-0.30

Table A.13.: *Hydrolysis of formic and acetic acid esters (60° C, pH 3) for 1, 2, 3, 4 and 7 d of sample G004541.*

G001980; Formic acid; M: 46 g/mol, T: 45°C = 318.15 K					
t [s]	86400	172800	259200	345600	604800
bound formic acid (ester$_0$) [mg/gTOC]	7.36	7.36	7.36	7.36	7.36
hydrol. formic acid [mg/gTOC]	0.3	0.21	0.2	no data	0.01
resid. formic acid (ester) [mg/gTOC]	7.06	7.15	7.16	no data	7.35
resid. formic acid (ester) [mmol/gSed]	0.153	0.155	0.156	no data	0.160
ln(ester/ester$_0$)	-0.04	-0.03	-0.03	no data	0.00
G001980; Acetic acid; M: 60 g/mol, T: 45°C = 318.15 K					
bound acetic acid (ester$_0$) [mg/gTOC]	2.24	2.24	2.24	2.24	2.24
hydrol. acetic acid [mg/gTOC]	0.48	0.26	0.27	no data	0.04
resid. acetic acid (ester) [mg/gTOC]	1.76	1.98	1.97	no data	2.20
resid. acetic acid (ester) [mmol/gTOC]	0.029	0.033	0.033	no data	0.037
ln(ester/ester$_0$)	-0.24	-0.11	-0.11	no data	0.00

Table A.14.: *Hydrolysis of formic and acetic acid esters (45° C, pH 3) for 1, 2, 3, 4 and 7 d of sample G001980.*

G001996; Formic acid; M: 46 g/mol, T: 45°C = 318.15 K					
t [s]	86400	172800	259200	345600	604800
bound formic acid (ester$_0$) [mg/gTOC]	7	7	7	7	7
hydrol. formic acid [mg/gTOC]	0.19	0.12	0.01	0.19	0.01
resid. formic acid (ester) [mg/gTOC]	6.81	6.88	6.99	6.81	6.99
resid. formic acid (ester) [mmol/gTOC]	0.148	0.150	0.152	0.148	0.152
ln(ester/ester$_0$)	-0.03	-0.02	0.00	-0.03	0.00
G001996; Acetic acid; M: 60 g/mol, T: 45°C = 318.15 K					
bound acetic acid (ester$_0$) [mg/gTOC]	1.63	1.63	1.63	1.63	1.63
hydrol. acetic acid [mg/gTOC]	0.28	0.16	0.01	0.18	0.03
resid. acetic acid (ester) [mg/gTOC]	1.35	1.47	1.62	1.45	1.60
resid. acetic acid (ester) [mmol/gTOC]	0.023	0.025	0.027	0.024	0.027
ln(ester/ester$_0$)	-0.16	-0.08	0.00	-0.12	0.00

Table A.15.: *Hydrolysis of formic and acetic acid esters (60°C, pH 3) for 1, 2, 3, 4 and 7 d of sample G001996.*

G001989; Formic acid; M: 46 g/mol, T: 45°C = 318.15 K					
t [s]	86400	172800	259200	345600	604800
bound formic acid (ester$_0$) [mg/gTOC]	3.49	3.49	3.49	3.49	3.49
hydrol. formic acid [mg/gTOC]	0.08	0.07	0.07	no data	0.1
resid. formic acid (ester) [mg/gTOC]	3.41	3.42	3.42	no data	3.39
resid. formic acid (ester) [mmol/gTOC]	0.074	0.074	0.074	no data	0.074
ln(ester/ester$_0$)	-0.03	-0.03	-0.03	no data	-0.03
G001989; Acetic acid; M: 60 g/mol, T: 45°C = 318.15 K					
bound acetic acid (ester$_0$) [mg/gTOC]	0.89	0.89	0.89	0.89	0.89
hydrol. acetic acid [mg/gTOC]	0.21	0.11	0.12	no data	0.16
resid. acetic acid (ester) [mg/gTOC]	0.68	0.78	0.77	no data	0.73
resid. acetic acid (ester) [mmol/gTOC]	0.011	0.013	0.013	no data	0.012
ln(ester/ester$_0$)	-0.31	-0.14	-0.14	no data	-0.22

Table A.16.: *Hydrolysis of formic and acetic acid esters (45°C, pH 3) for 1, 2, 3, 4 and 7 d of sample G001989.*

A. Tables

	Sample G004541			
T [°C]	90	75	60	45
T [K]	363.15	348.15	333.15	318.15
1/T [K^{-1}]	0.002753683	0.002872325	0.003001651	0.003143171
Formate				
k'(pH 7)	$3.48 \cdot 10^{-11}$	$2.05 \cdot 10^{-11}$	$2.13 \cdot 10^{-11}$	$1.13 \cdot 10^{-11}$
ln k'	-24.0814037	-24.6105962	-24.5723140	-25.2062184
Acetate				
k'(pH 7)	$2.97 \cdot 10^{-10}$	$1.05 \cdot 10^{-10}$	$6.80 \cdot 10^{-11}$	$3.00 \cdot 10^{-11}$
ln k'	-21.9372890	-22.9770608	-23.4115134	-24.2298237

Table A.17.: *Data calculation for Arrhenius plot of sample G004541.*

	Sample G001980			
T [°C]	90	75	60	45
T [K]	363.15	348.15	333.15	318.15
1/T [K^{-1}]	0.002753683	0.002872325	0.003001651	0.003143171
Formate				
k'(pH 7)	$5.84 \cdot 10^{-11}$	$3.51 \cdot 10^{-11}$	no data	($-7.25 \cdot 10^{-8}$)
ln k'	-24.7919427	-25.2607774	-	-
Acetate				
k'(pH 7)	$2.98 \cdot 10^{-10}$	$3.13 \cdot 10^{-10}$	no data	($-3.68 \cdot 10^{-7}$)
ln k'	-23.420376	-24.1528627	-	-

Table A.18.: *Data calculation for Arrhenius plot of sample G001980.*

	Sample G001996			
T [°C]	90	75	60	45
T [K]	363.15	348.15	333.15	318.15
1/T [K^{-1}]	0.002753683	0.002872325	0.003001651	0.003143171
Formate				
k'(pH 7)	$9.77 \cdot 10^{-12}$	$8.16 \cdot 10^{-12}$	$9.47 \cdot 10^{-12}$	($-3.80 \cdot 10^{-12}$)
ln k'	-25.3517047	-25.5317770	-25.3828922	-
Acetate				
k'(pH 7)	$3.23 \cdot 10^{-10}$	$9.40 \cdot 10^{-11}$	$4.69 \cdot 10^{-11}$	($-2.29 \cdot 10^{-11}$)
ln k'	-21.8533680	-23.0877263	-23.7270303	-

Table A.19.: *Data calculation for Arrhenius plot of sample G001996.*

	Sample G001989			
T [°C]	90	75	60	45
T [K]	363.15	348.15	333.15	318.15
1/T [K^{-1}]	0.002753683	0.002872325	0.003001651	0.003143171
Formate				
k'(pH 7)	$3.37 \cdot 10^{-13}$	($-1.15 \cdot 10^{-12}$)	$2.66 \cdot 10^{-12}$	$1.59 \cdot 10^{-12}$
ln k'	-28.7186935	-	-26.6526950	-27.1672871
Acetate				
k'(pH 7)	$2.44 \cdot 10^{-10}$	$1.72 \cdot 10^{-10}$	$3.21 \cdot 10^{-11}$	($-2.29 \cdot 10^{-12}$)
ln k'	-22.1338529	-22.4835266	-24.1621651	-

Table A.20.: *Data calculation for Arrhenius plot of sample G001989.*

B. Mass spectra

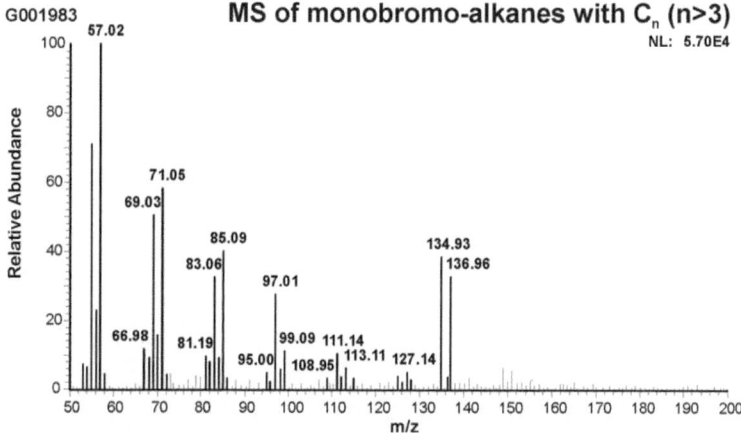

Figure B.1.: *MS of all aliphatic monobro-malkanes (m/z=135: $C_4H_8Br_2^+$).*

B. Mass spectra

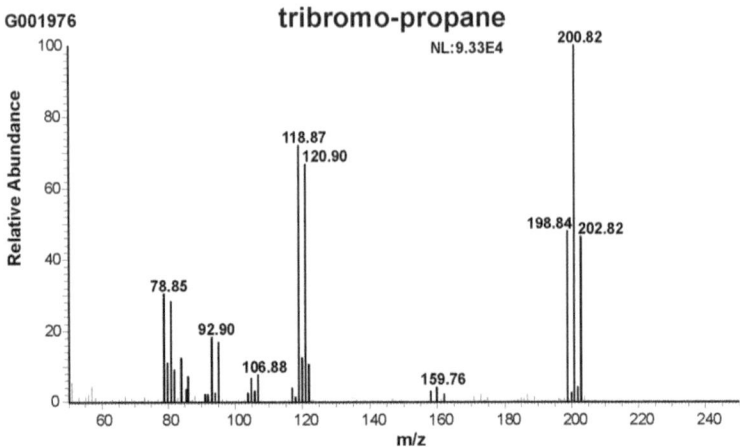

Figure B.2.: *MS of tribromo-propane, identified from mainlib-libary. The molecular ion at m/z 280.8 is not visible in the spectrum due to rapid Br elimination.*

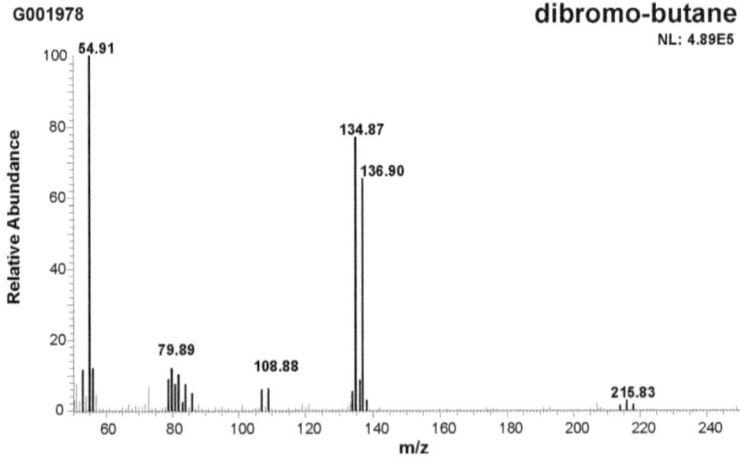

Figure B.3.: *MS of dibromo-butane, identified from mainlib-libary.*

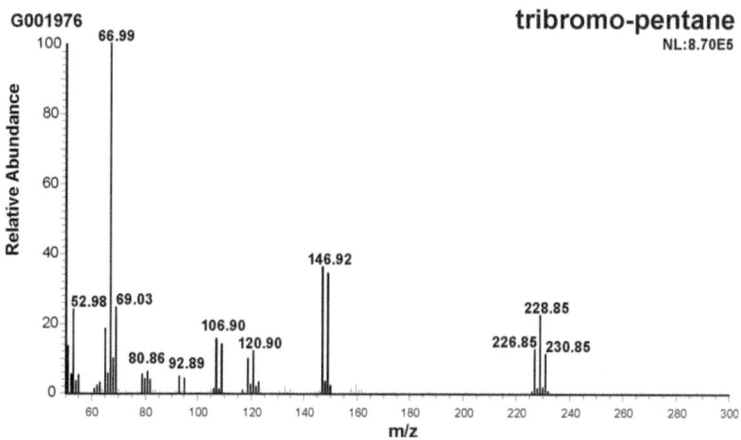

Figure B.4.: *MS of tribromo-propane, identified from mainlib-libary. The molecular ion at m/z 308.8 is not visible in the spectrum due to rapid Br elimination.*

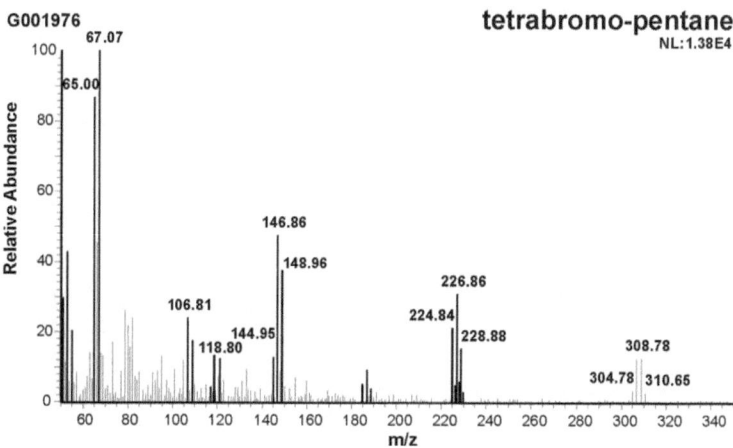

Figure B.5.: *MS of tetrabromo-propane, identified from Wollenweber et al. (2006).*

B. Mass spectra

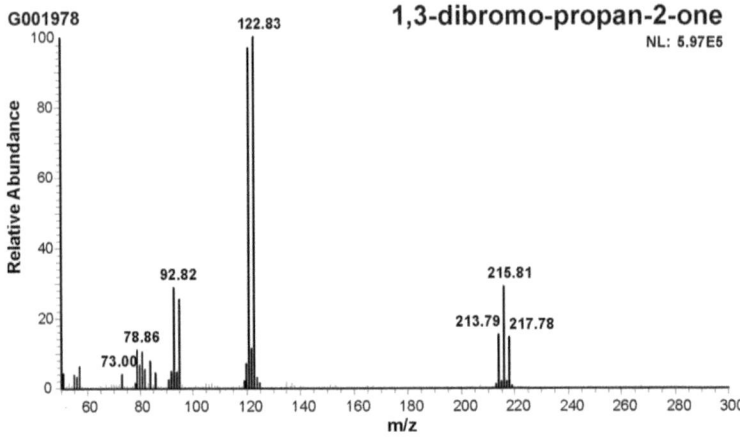

Figure B.6.: *MS of 1,3-dibromo-propan-2-one, identified from mainlib-libary.*

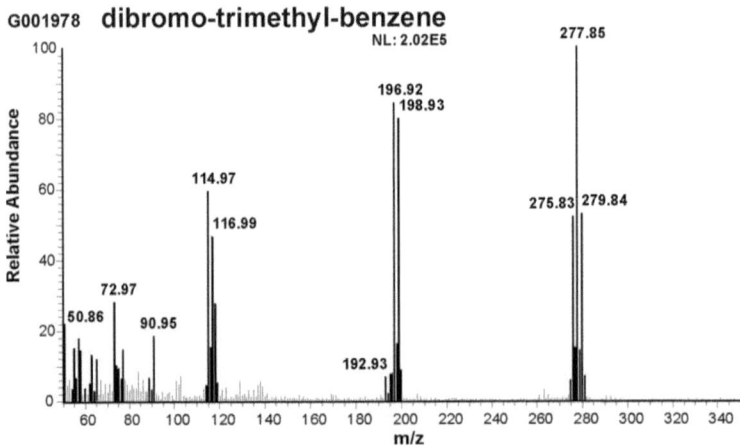

Figure B.7.: *MS of dibromo-trimethyl-benzene, identified from mainlib-libary.*

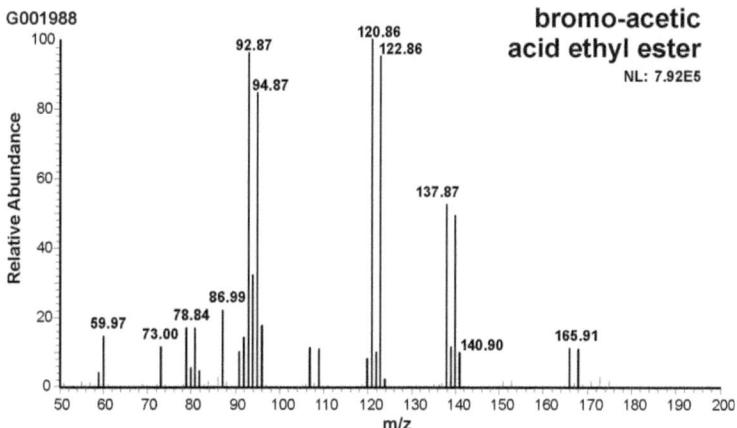

Figure B.8.: *MS of bromo-acetic-acid-ethyl-ester, identified from mainlib-libary.*

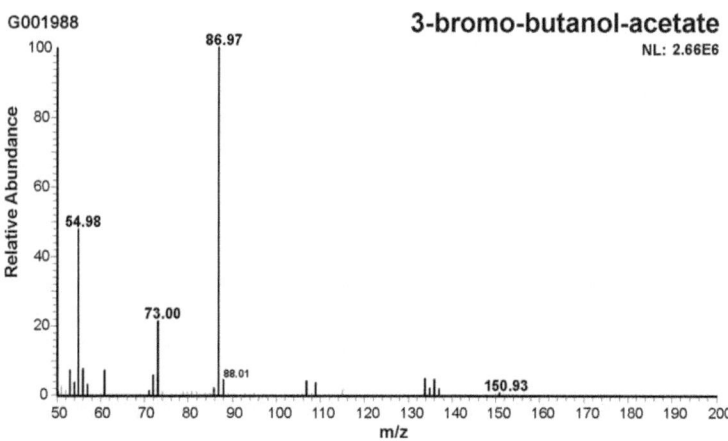

Figure B.9.: *MS of 3-bromo-butanole-acetate, identified from mainlib-libary.*

C. Publications

In the scope of this thesis the following articles and conference contributions were published.

Articles (ISI):

Glombitza, C., Mangelsdorf, K., Horsfield, B. (2011). Structural insights from boron tribromide ether cleavage into lignites and low maturity coals from the New Zealand Coal Band. Organic Geochemistry, 42(3):228-236.

Glombitza, C., Mangelsdorf, K., Horsfield, B. (2009). Maturation related changes in the distribution of ester bound fatty acids and alcohols in a coal series from the New Zealand Coal Band covering diagenetic to catagenetic coalification levels. Organic Geochemistry, 40(10):1063-1073.

Glombitza, C., Mangelsdorf, K., Horsfield, B. (2009). A novel procedure to detect low molecular weight compounds released by alkaline ester cleavage from low maturity coals to assess its feedstock potential for deep microbial life. Organic Geochemistry, 40(2):175-183.

C. Publications

Presentations on conferences:

Mangelsdorf, K., Glombitza, C., Vieth, A., Vu Thi Ahn, T., Kallmeyer, J., Zink, K.-G., Sykes, R., Fry, J. and Horsfield, B. (2011). Microbial communities associated to deep subsurface coal layers: A review of the DEBITS project. 25th International Meeting on Organic Geochemistry - IMOG (Interlaken, Switzerland, 18.09.-23.09.2011), Book of Abstracts, 98.

Glombitza, C., Mangelsdorf, K., Vieth, A., Sykes, R., K., Horsfield, B. (2009). Low mature coals as potential feedstock for deep microbial life. 24th International Meeting on Organic Geochemistry - IMOG (Bremen, Germany, 06.09.-11.09.2009), Book of Abstracts, 103.

Glombitza, C., Mangelsdorf, K., Horsfield, B.(2008). Alkaline ester cleavage to examine the feedstock potential of low mature coals for deep microbial populations. European Geosciences Union General Assembly - EGU (Vienna, Austria, 13.04.-18.04.2008), Geophysical Research Abstracts, 10, EGU2008-A-07643.

Glombitza, C., Mangelsdorf, K., Horsfield, B. (2007). Investigation of the structural composition of sedimentary organic matter to assess its feedstock potential for deep microbial populations. 23rd International Meeting on Organic Geochemistry - IMOG (Torquay, Devon, United Kingdom, 09.09.-14.09.2007), Book of Abstracts, 439-440.

Acknowledgements

First of all my thanks go out to my supervisor Dr. Kai Mangelsdorf for uncountable inspiring discussions and his excellent supervision, without this, the present work would not have been possible, and to Prof.Dr. Brian Horsfield for providing the themes and topics under investigation and the place in his organic geochemistry group and with this, for giving me the opportunity to receive an insight into this intriguing and challenging scientific discipline. Further, i'd like to thank Prof.Dr. Philippe Schaeffer (CNRS, Strassbourg) and Dr. Jan Schwarzbauer (RWTH, Aachen) for sharing knowledge and providing tips on chemical degradation procedures. The Helmholtz Association is gratefully acknowleged for the financial support of this work.

Special thanks go out to the technical staff at the organic geochemistry group at GFZ, namely Cornelia Karger for GC-MS measurements, Kristin Günter for IC measurements and Ferdinand Perssen for pyrolysis GC-MS and pyrolysis GC-FID measurements as well as Anke Kaminsky for support in laboratory work. The guest student Jens Müller is thanked for the good lab work during his internship. I'd like to thank and salute my colleagues and friends Katja Theuerkorn, Alex Vetter and Nick Mahlstedt for sharing all the good and bad times. Furthermore, I'd like to thank all fellows at the OG group for providing a good working atmosphere.

Ich danke meiner Familie, die mich immer mit Interesse und Hilfsbereitschaft begleitet haben. Mein Dank gilt ebenso meinen Freunden, die mir reichlich Abwechslung verschafft haben und mein Leben stets vielseitig halten.

Die VDM Verlagsservicegesellschaft sucht für wissenschaftliche Verlage abgeschlossene und herausragende

Dissertationen, Habilitationen, Diplomarbeiten, Master Theses, Magisterarbeiten usw.

für die kostenlose Publikation als Fachbuch.

Sie verfügen über eine Arbeit, die hohen inhaltlichen und formalen Ansprüchen genügt, und haben Interesse an einer honorarvergüteten Publikation?

Dann senden Sie bitte erste Informationen über sich und Ihre Arbeit per Email an *info@vdm-vsg.de*.

Sie erhalten kurzfristig unser Feedback!

VDM Verlagsservicegesellschaft mbH
Dudweiler Landstr. 99   Telefon +49 681 3720 174
D - 66123 Saarbrücken   Fax   +49 681 3720 1749
www.vdm-vsg.de

Die VDM Verlagsservicegesellschaft mbH vertritt

Printed by Books on Demand GmbH, Norderstedt / Germany